T0211096

Virus dynamics

Virus dynamics

Mathematical principles of immunology and virology

Martin A. Nowak

Robert M. May

OXFORD

UNIVERSITY PRESS

*This book has been printed digitally and produced in a standard specification
in order to ensure its continuing availability*

OXFORD

UNIVERSITY PRESS

Great Clarendon Street, Oxford OX2 6DP

Oxford University Press is a department of the University of Oxford.
It furthers the University's objective of excellence in research, scholarship,
and education by publishing worldwide in

Oxford New York

Auckland   Cape Town   Dar es Salaam   Hong Kong   Karachi
Kuala Lumpur   Madrid   Melbourne   Mexico City   Nairobi
New Delhi   Shanghai   Taipei   Toronto
With offices in
Argentina   Austria   Brazil   Chile   Czech Republic   France   Greece
Guatemala   Hungary   Italy   Japan   South Korea   Poland   Portugal
Singapore   Switzerland   Thailand   Turkey   Ukraine   Vietnam

Oxford is a registered trade mark of Oxford University Press
in the UK and in certain other countries

Published in the United States
by Oxford University Press Inc., New York

Oxford is a registered trade mark of Oxford University Press
in the UK and in certain other countries

Published in the United States
by Oxford University Press Inc., New York

ISBN 0-19-850417-9

PREFACE

We know, down to some of the tiniest details, the molecular structure of the human immunodeficiency virus (HIV). Deriving from a variety of technological and conceptual advances in recent years, this accomplishment represents a triumph of human ingenuity, which in some ways defies any intuitive grasp. But, despite such remarkable advances at the level of understanding individual viruses and individual cells of the immune system—and their interaction—we still have no agreed understanding of why there is so long and variable an interval between infection with HIV and the onset of AIDS. Such ignorance of the ultimate course and variability of pathogenesis impedes efforts towards developing effective long-term therapies or preventive vaccines.

We believe part of the answer to these unresolved questions lies in a better understanding of how entire populations of viruses like HIV interact with entire populations of different kinds of immune system cells. Such population models must, of course, be grounded on empirical facts about the interactions between individual viruses and cells. And any emerging conclusions need to be tested against data from laboratory experiments and, even more important, the clinical course of infection in patients. Understanding the population dynamics of the engagement between a viral infection and the immune system cells is thus only one small part of a fully mature discipline of immunology, but it will often be a crucial part.

That is the central premise, and indeed the motive, for this book. We aim to set out ideas about how populations of viruses and populations of immune system cells may interact in various circumstances. These ideas are most precisely expressed as mathematical models; mathematics is no more, but no less, than a way of thinking clearly. The very simplest of these mathematical models are nonlinear, which immediately implies that simple assumptions can produce exceedingly complex dynamics (with the corollary that very complex immunological happenings in individual patients can be produced by very simple underlying mechanisms). Throughout the book, we aim to present the ideas and assumptions clearly and explicitly, and wherever possible intuitively, emphasising the empirical underpinning, and suggesting possible empirical tests along with implications for treatments or vaccine development.

Our collaboration began in 1989. One of us (MAN) had just completed a PhD with Karl Sigmund and Peter Schuster in Vienna, and had been awarded an Erwin Schroedinger Fellowship for postdoctoral study abroad. His interest in questions raised at that time in a *Nature* News and Views piece (on the evolution of altruism) by RMM motivated him to come to Oxford. RMM had just completed his book on *Infectious Diseases of Humans* with Roy Anderson. This book derived from an earlier interest in the extent to which infectious diseases may affect the numerical abundance or geographical distribution of non-human animal populations, but was focussed on the transmission and control of viral, bacterial, protozoan, and helminth infections in human populations. In

Chapter 3 of Anderson and May's book, by way of background, aspects of the nonlinear dynamics of the immune system were explored, using very simple, cartoon-like models. This work was itself stimulated by the pioneering efforts of Bell, Hoffman, Jerne, and others, who earlier asked the kinds of questions that a theoretical ecologist or population biologist might ask of such a system of "interacting species".

It therefore seemed natural for the two of us to work on new approaches to the above-mentioned "HIV/AIDS problem". We began with the very simplest model incorporating the essential features of the interaction between HIV and the immune system: the Basic Model of Chapter 12. The strange dynamical properties of this model, with its counter-intuitive "diversity threshold", encouraged us. This led to collaborations with a variety of experimental and clinical groups, in pursuit of data to constrain the models and of tests of emerging conclusions. In turn, this led to the models' acquiring an encrustation of detail and complexity, and maybe even a patina of realism. Also the general approach has been extended to other, differently-acting viral infections, particularly hepatitis B (HBV). The resulting papers—some short, some long, but with little overlap among them—are scattered among generalist and various specialist journals (biomedical, ecological, and theoretical biological). Moreover, as the work has progressed, our views have developed and changed somewhat.

Hence this book. It aims to give a broad impression of the still-nascent discipline of theoretical immunology, in the sense of the dynamical properties of interacting populations of viruses and immune system cells, with special emphasis on HIV/AIDS. The book makes no claim to be encyclopaedic. Rather it is a personal, not to say idiosyncratic or even narcissistic, view of a still developing field. We try to convey a sense of how feeling this particular part of the immunology elephant relates to the subject as a whole, of the useful insights that can emerge, and of some of the limitations of the approach.

Most chapters begin with an overview, expressed as far as possible in intuitive terms; the biological basis for the mathematical models under discussion, and the lessons to be drawn from them. The main text of each chapter then describes its empirical and theoretical content in more detail, explaining and justifying the conclusions about those aspects of the immune systems' dynamics, which are being studied. In some cases, more technical details of the mathematical developments are banished to Appendices; in other cases, where the details are less interesting, the reader is referred to original papers. Most chapters end with a brief summary. A quick overview of the potential value, and of the shortcomings, of this general approach to understanding aspects of the immune system can, indeed, be gained by skipping through the chapters, reading only the introduction and summary of each. But we would like you to do more!

The opening chapter sets the stage. It outlines the basic workings of the immune systems in response to viral challenge, showing how biological statements translate into nonlinear equations for the constituent populations of viruses and cells. It explains our intention to keep our mathematical models, in Einstein's words, "as simple as possible, but not more so".

Chapter 2 gives the biology and some social history of HIV/AIDS. The next three chapters focus on viral dynamics: the Basic Model of virus dynamics is introduced in Chapter 3; it is used to explore the dynamics of anti-viral therapy against HIV in

Chapter 4; and the dynamics of HBV are explored in Chapter 5. Returning to HIV, Chapters 6 and 7 proceed to the full set of interacting viral and immune cell populations, combining theory and data to illuminate the dynamics of immune responses (Chapter 6), and examining the rates at which immune responses can eliminate infected cells (Chapter 7). We next introduce the notion of a quasi-species (Chapter 8), briefly examine factors determining the frequency of drug-resistant viral strains before anti-viral therapy is initiated (Chapter 9), and explore the subsequent rate of emergence of such drug resistance (Chapters 10 and 11). Progressively more elaborate, and correspondingly more realistic, models with antigenic variation are considered in Chapters 12, 13, and 14.

The final Chapter 15 is about "everything we know so far and beyond". It guesses at what are the next empirical and theoretical steps and the next exciting questions. Such chapters often turn out, in retrospect, to be embarrassingly wrong, but we are both risk-takers.

We emphasise that the present book only scratches the surface of the emerging field of theoretical immunology. More precisely, it scratches only one facet of the surface, namely viral-cell population interactions. Other major questions in immunological theory have received even less attention as yet.

For example, interactions between viruses or other infectious agents and immune system cells must play a central role in any basic understanding of the evolution of virulence. Recent years have seen considerable advances in this area, but they come mainly from phenomenological analyses of how the Darwin/Fisher "fitness" or "basic reproductive number" of a virus is influenced by its transmissibility (producing new infections) and virulence (which affects the rate of recovery, or death from the infection, of the host). Such studies are phenomenological in the sense that functional relations between transmissibility and virulence are either inferred (usually very roughly) from data, or (more usually) invented for illustrative purposes. It would be good to see more work in which both transmissibility and virulence were derived, at least to some degree, from explicit models of the interplay between populations of viruses and populations of immune system cells. In early efforts in this direction, we have found interesting differences in the general trends in evolution of virulence between systems with "coinfection" (where a host can harbour two or more coexisting strains of the virus) and those with "superinfection" (where the most virulent strain takes over). But this is another story. (See Nowak and May 1994 and May and Nowak 1994, 1995.) It is a part of the immunological forest, at the boundary between evolutionary history and medical implications that deserves more attention.

In short, our book is a personal view of one emerging, and potentially highly useful, area of theoretical immunology. You could call it, in analogy with the subareas of theoretical ecology and evolution, the population dynamics of the immune system. Although you will not meet them again in this book, there are other important general areas—the ecosystem and evolutionary dynamical aspects of theoretical immunology—that cry out for attention and new ideas.

Finally, the list of people to whom we are indebted is too long to include in its entirety. We must, howeyer, single out Roy Anderson, Charles Bangham, Sebastian Bonhoeffer, Angela McLean, Andrew McMichael, George Shaw, Jeff Lifson, Dominik

Wodarz, Rodney Phillips, Ruy Ribeiro, Karl Sigmund, Alun Lloyd, Paul Klenerman, and Roland Regoes among the many who have provided much stimulus, along with good advice. This book could not have been written without the help of Donna Welton and Joanna Lacey. MAN is grateful to the Royal Society, the Wellcome Trust, Wolfson and Keble Colleges at Oxford, and the Institute for Advanced Study in Princeton for support at various stages. RMM is supported by his Royal Society Research Professorship. In a previous book, RMM cryptically thanked his dog for emotional support. More conventionally, but no less sincerely, this time we both thank our families.

Princeton, Oxford   Martin A. Nowak
September 2000   Robert M. May

CONTENTS

1

INTRODUCTION: VIRUSES, IMMUNITY, EQUATIONS

1.1 Viruses

Viruses populate the world between the living and the non-living. They themselves are not capable of reproduction, but if put into the right environment they can manipulate a cell to generate numerous copies of themselves. 'Reproduce me!' is the essence of the virus, the message that the viral genome carries into the headquarter of a cell—the nucleus. The viral genome manages to attract the attention of the workers in the cell, which are the various enzymes capable of copying and interpreting genetic information. These workers will read the viral message and follow its instruction; they will produce viral proteins and more copies of the viral genome. The cell may devote all its resources to produce new virus particles and die after everything has been turned into virus. A swarm of new 'Reproduce me!' messages is then leaving the cell, searching for new targets.

But viruses can also have more subtle programs than this. They may tell the host cell to reproduce them, but only at a slow rate not endangering the survival of the cell. They may enter a cell, insert their genetic material into its genome and be very quiet for a long time. Under specific conditions they may become reactivated and demand their reproduction. Other viruses once inserted into the genome of the cell may induce the cell to divide thereby producing two *infected* daughter cells. Such viruses may drive their host cell into uncontrolled multiplication, and thereby cause cancer.

'Wee animalcule' was Antony van Leeuwenhoek's expression for the living creatures which populated the world under his brilliant microscope. Leeuwenhoek lived in Holland, in the seventeenth century. His specialty was to grind the best lenses of his time. They offered him a 300-fold magnification, just sufficient to detect bacteria and other unicellular organisms. In a series of papers to the Royal Society of London, Leeuwenhoek described in detail his newly discovered 'microbial' world, which was richly equipped with bizarre forms and movements revealing a great biological diversity previously completely unknown to humans.

But lens grinding was some art in those days and few people had microscopes as good as Leeuwenhoek's. Carolus Linnaeus knew only six species of microbes which he classified in 1767 under the appropriate name 'Chaos'.

The nineteenth century microbiologists were puzzled by the question whether these 'wee animalcule' can arise by 'spontaneous generation' in decaying biological material or only by duplication, that is, reproduction of themselves. Louis Pasteur performed the decisive experiment and showed that sterilized medium remained free of microbes in swan-neck flasks, into which air but no microbes could enter. Spontaneous generation did not occur.

Scientists also began to study the diseases associated with microbes. Epidemics of contagious diseases have tormented humans for millennia, but the nature of the infection process was not understood. Ancient Greeks believed that bad vapours transmitted plague. In the nineteenth century it was suspected that the culprits of various infectious diseases of plants, animals and humans were to be found in the microbial world.

Robert Koch outlined a scheme how to prove that a certain microbe was the agent of a certain disease. Koch demanded that (i) the organism must be found regularly associated with the disease, (ii) the organism must be isolated in pure culture, (iii) inoculation of such a culture into the host must cause disease, and (iv) the organism must be recovered once again from the inoculated host. Koch's postulates helped to define the infectious agents for a variety of animal, human and plant diseases.

In 1892, the Russian biologist Dimitrii Ivanovsky studied tobacco mosaic disease of plants. The infectious agent was able to replicate on the leaves of plants, but not in media that would normally support bacterial growth. Furthermore, it was apparently too small to be seen with a microscope. Ivanovsky showed that the infectious agent could pass through special filters that had pores too small for any bacteria. Something smaller than any known organism had to be at work here. Ivanovsky had discovered the first virus, the first member of the submicroscopic world.

The discovery of other viruses followed rapidly. In 1900, Walter Reed described the first human viral disease: yellow fever. Today hundreds of viruses are known, infecting humans, plants, animals and even bacteria. Viruses are usually classified according to their host, chemical composition and structure. Many important and devastating human diseases are caused by viruses. In 1919, influenza A virus killed 20 million people. Currently, 300 million people are believed to be infected with the hepatitis B virus (HBV) and 25% of them may die from liver cancer caused by the virus. The worldwide pandemic of HIV has at the turn of the millennium reached about 50 million people: 15 million have already died from AIDS, 35 million are currently infected. In certain urban regions of Africa, up to 40% of the people carry the virus, and it is estimated that this infection alone could reverse population growth in Africa. In the United States, AIDS was the leading cause of death in young men before the introduction of triple-drug therapy.

Viruses are as old as life itself. Even the simplest living cells (bacteria) are subject to viral infections. It seems that any genetic system, which has the ability of reproducing genetic information, can be exploited by viruses. Even theories for the origin of life have to come up with mechanisms how early replicators can protect themselves against exploitation by viral parasites.

Viruses contain their genetic information either in form of DNA or RNA. Most viruses have a protein coat protecting this genetic material. Some viruses also have an envelope consisting of lipids and additional proteins.

All viruses need to infect cells for their reproduction. The life-cycle of viruses can be subdivided into eight key events:

1. The virus attaches itself to the host cell.
2. The virus (or at least its genetic material) invades the cell.
3. The genetic material of the virus is uncoated.
4. Viral proteins are produced which start manipulating the host cell.

5. The viral genome is multiplied by the machinery of the host cell.
6. Further viral proteins are produced which will be used for the coat and possibly envelope of the new virus particles.
7. The new virus particles are assembled.
8. The new virus particles are released from the host cell.

If viruses enter the human body, they normally do not replicate unabated, but are opposed by immune responses. Evolution has equipped all vertebrates with a complex and ingenious system to combat viruses and other infectious diseases. A rough outline of how the immune system works will be the subject of the next section.

1.2 Immunity

One microlitre of human blood contains about 2500 lymphocytes, small cells which constitute the backbone of the immune system. But only 2% of these cells reside in the blood, the rest can be found distributed throughout the body, in the various organs of the immune system, such as the lymphatic tissue, the spleen, the thymus and the bone marrow. In total, there are about 10^{12} lymphocytes in an adult human. Lymphocytes can be subdivided into B and T cells.

1.2.1 *B cells*

B cells are produced in the bone marrow. They carry antibody molecules on their cell surface. An ingenious genetic mechanism ensures that antibody molecules are extremely diverse: the subunits of the antibody protein can be combined in many ways to give an almost indefinite repertoire of different antibody molecules. Essentially for any foreign molecule entering the human body, there is a specific antibody molecule that can bind to it like a lock and key mechanism.

Suppose a foreign pathogen, for example a virus, enters the human body. Viral proteins will encounter B cells. Most of the B cells will not have antibody receptors of the correct specificity, but some B cells will be able to bind to some viral proteins. These specific B cells will become activated and will start to divide thereby increasing in number. This process is called 'clonal selection', because the best fitting B-cell clone is selected to multiply. (A clone is a population of cells deriving from the same ancestor.)

During the multiplication of the B cells additional variation of the antibody molecule is introduced. This 'somatic mutation' generates B cells with antibody molecules with slightly different shapes. Some of these antibodies may provide a better fit with the viral protein than the original antibodies. The B cells producing the improved antibodies will receive a stronger activation signal from interaction with the virus and will multiply at a faster rate. They will outcompete the other B cells. This process is called 'affinity maturation'. After some time the B-cell response will have optimum specificity for the invading viral particles.

B cells can release their antibody molecules into the blood. These antibodies then bind to the virus particle and mark it as a foreign structure for elimination by other cells of the immune system. Immune cells, called 'macrophages' have the task of digesting structures that are bound to antibodies. In addition there is a system of blood enzymes, called 'complement system' which binds to antibody coated structures and removes them.

It is also possible that the antibody coating of a virus particle is so dense that this virus particle is prevented from invading its target cells. Antibodies that can achieve this goal are called 'neutralizing'.

1.2.2   T cells

T cells are the other key players of the human immune system. They are generated in the thymus. There are CD8 positive T cells, which are killer cells, and CD4 positive T cells, which are helper cells. CD8 and CD4 refer to some proteins on the surface of these cells. Proteins on the surface of cells are called receptors, because they often bind to other molecules. There are many CD-something receptors with important functions, and immunologists classify the cells of the immune system according to their receptors.

The role of the immune system is to fight off invasion by foreign pathogens. In order to do this the immune system has to be able to distinguish between 'self' and 'non-self'. The responsibility for this key feature lies with T cells. In the following we will outline the basic principle of self–non-self distinction.

CD8 cells have the ability to recognize and eliminate cells that are infected by virus. Essentially every cell of the human body has a complicated machinery that cuts up proteins inside the cell and presents parts of these proteins (peptides) on the surface of the cell. These peptides are bound to major histocompatibility complex class one (MHC-1) molecules. The peptide–MHC complex on the surface of the cell is seen by CD8 cells. CD8 cells carry on their surface a protein called T-cell receptor, which is designed to bind to MHC–peptide complexes. This interaction is highly specific: the T-cell receptor and the peptide–MHC complex will only bind if they fit together like lock and key.

MHC molecules have a specificity for certain peptides but they bind equally well to peptides that derive from cellular or viral proteins. The distinction between self and non-self is on the level of the CD8 cell. During their generation in the thymus only those CD8 cells are permitted to enter into the blood which do not bind to MHC molecules in association with self-peptides, which derive from normal cellular proteins. This process is called negative selection. It is essentially based on the assumption that all peptides present in the thymus are self-peptides and therefore any T cell that would react against such a peptide–MHC complex has to be eliminated. Interestingly, if a virus would manage to infect the thymus during T-cell development then the immune system might consider this virus as self.

There is also positive selection during T-cell development. The T-cell receptor is generated similar to the antibody molecule. An ingenious genetic mechanism combines various parts of the T-cell receptor to produce a large variety of different T-cell receptor molecules. There is an enormous combinatorial repertoire. In the thymus only those T cells are allowed to survive that have at least some capacity to bind to the existing MHC molecules. T cells that do not bind to any MHC molecules would be useless.

Thus, positive and negative selection in the thymus work in concert to produce a T-cell population that has the ability to bind to the correct MHC molecules in a given individual, but do not cause immune responses to self-peptides.

If a cell becomes infected by a virus, the virus induces the cell to produce viral proteins. These proteins are cut into pieces and presented as peptide–MHC complexes on the surface of the cell. If a CD8 cell with the correct T-cell receptor finds an infected cell, it will bind to the peptide–MHC complex. The CD8 cell will become activated and start to produce chemicals that kill the target cell. These 'lethal hits' can be observed under a microscope: T cells become agitated when they come into the vicinity of an infected cell, it almost seems as if they can 'smell' virus. They will attach themselves to the cell and move over its surface. If their receptor finds the correct peptide–MHC complex they will stay longer (otherwise they may fall off again). A large number of T-cell receptors will bind to peptide–MHC complexes, which acts as an activation signal for chemical processes inside the cell. The T cell will release substances that can make holes into the cell membrane of the infected cell. After some time the infected cell will 'explode' and when the 'smoke' settles the T cell will appear again and move away ready for further action.

The CD8 cell after having eliminated the infected cell will remain activated and look for further infected cells in the immediate neighbourhood. It may also divide and give rise to two daughter cells of the same specificity ready to kill more virus infected cells.

CD8 cells also release chemicals that have the ability to prevent virus replication in infected cells. They can also activate defence mechanisms inside the cell or prevent viruses from infecting cells.

CD4 cells do not themselves kill infected cells or remove virus particles, but they help both B cells and CD8 cells to mount effective immune responses. CD4 cells have a T-cell receptor similar to CD8 cells, but they recognize peptides in association with MHC class 2 molecules. MHC-2 molecules are only on the surface of specific cell types in the human body, such as B cells and macrophages.

When B cells encounter a virus particle and their antibody receptor can bind to a specific part of this virus, they internalize the virus, digest it inside and present virus peptides in association with MHC-2 molecules on their surface. CD4 cells will now check this peptide–MHC complex, and if they can bind to it they will signal to the B cell to be alert and to divide. This is an important mechanism helping the B-cell response to distinguish between self and non-self.

Similarly macrophages digest foreign or suspicious material inside the body and present peptides with MHC-2 to CD4 cells. If the material is non-self then the CD4 cell will activate the macrophage to divide and become more aggressive.

CD4 cells release substances that activate immune responses. In some sense, if they have been presented with foreign peptides, they sound a warning signal and activate many different components of the immune system some of which are specific (such as CD8 cells and B cells) while others are unspecific (such as macrophages). The distinction between self and non-self, however, is never absolutely perfect. A highly activated immune system can also react against uninfected cells and cause inflammation which might be disadvantageous. The important goal is to stop the infection, to prevent the invading agent from killing the host.

The immune system of vertebrates with its B and T cells has been shaped over millions of years of evolution. There was continuous exposure to infectious agents creating selection for hosts that would survive given infections. The immune system as we see it now is the product of these selection forces. Unfortunately, evolution also works for the microbes continuously selecting those mutants that can overcome the immune defence and establish infections. Most microbes have much shorter generation times than their hosts and reproduce in much greater numbers. Therefore from an evolutionary perspective they are much more flexible, which implies that they will be with us for ever.

1.3   Mathematical biology

> There is a shepherd and a flock of sheep. A man comes by and asks, 'If I guess the correct number of sheep in your flock can I have one?' The shepherd says, 'Please try.' The man says '83.' The shepherd is amazed; it is the correct number. The man picks up a sheep and walks away. The shepherd shouts, 'Hang on. If I guess your profession, can I have my sheep back?' The man says, 'Please try.' The shepherd says, 'Mathematical biologist.' The man is amazed, 'How did you know?' 'Because you picked up my dog.'

In 1202 the Italian mathematician Leonardo of Pisa (better known as Fibonacci) published a book which introduced the Hindu-Arabic decimal system to Western Europe. One of his examples was a problem of mathematical biology: 'How many pairs of rabbits can be produced from one pair, if every month each pair bears a new pair which from the second month on becomes productive?' It is assumed that 1 month elapses before the initial pair reproduces, that there are no deaths, and that each pair reproduces regularly. The number of adult rabbit pairs being present in consecutive months is then given by the Fibonacci sequence: 1, 1, 2, 3, 5, 8, 13, 21, ... (every term is the sum of its two predecessors). This work can be seen as the very beginning of mathematical biology, though one has to admit that the Fibonacci sequence exerts a greater fascination for number theorists than for reproductive biologists.

The greatest of all theories in biology is, of course, Darwinian evolution. As Theodosius Dobzhanski stated, nothing in biology makes sense if not seen in the context of evolution. The basic rules that explain the living world are mutation and selection. Mutation is random and generates biological variation. Selection chooses the survivors. We currently believe that this simple mechanism has given rise to every biological system starting from the simplest self-replicating molecule, to the first cell, to multicellular organisms, to the immune system and the nervous systems. Everything is at the most fundamental level explicable in terms of the interplay of mutation and selection.

In 1894, Oxford Zoologist Walter Weldon noted that Darwin's theory was intrinsically a mathematical theory and can only be tested with mathematical or statistical techniques. Weldon created a science called 'biometry', which is concerned with quantitative measurement in biology.

Charles Darwin was not a mathematician and is quoted for having regretted that he had no mathematical mind. According to some opinions, Darwin's lack of mathematical intuition prevented him from finding the genetic rules of inheritance in his plant experiments. Gregor Mendel, who performed similar experiments at about the same time, was to find the basic laws of inheritance.

Mendel was a student of Mathematics at the University of Vienna. He took many courses, but he did not receive a teacher's diploma because he failed Botany. Biology was not his strength. He became a monk to continue his studies. His famous pea crossings were conducted in the garden of an Augustine monastery in Brno (which is now a mathematics institute of the Czech academy of sciences). Guided by mathematical intuition and expectation of simple rules, Mendel discovered the foundation of genetics. Ronald Fisher referred to Mendel as 'a young mathematician whose statistical interest extended to physical and biological sciences'.

Interestingly, the laws of genetics were initially seen at variance with Darwin's theory of evolution: genetics did not appear to generate enough variability. This view was overcome in the 1920s and 1930s by the work of Ronald Fisher, J. B. S. Haldane and Sewall Wright, who are the founding fathers of mathematical biology. Their mathematical models show how mutation and selection work together to govern evolution.

For the beginnings of mathematical ecology, we have to go to Italy once again. During the first world war, there was not much fishing in the Adriatic sea. Some years afterwards the Italian biologist D'Ancona analysed the statistics of fish markets. He noted that during the war the proportion of predatory fish had increased. Why should war favour sharks? This was the question that D'Ancona asked his prospective father-in-law, Senator Vito Volterra, who was Professor of mathematical physics in Rome and one of the leading mathematicians of his time.

Volterra wrote down some differential equations and readily found the answer which is now known as Volterra's principle. It is worthwhile to look at a simple version of his model. Let us denote by x and y the population sizes of prey and predators, respectively. Suppose prey grow at the rate rx and they are killed by predators at the rate axy. Predators grow at a rate proportional to their own density times the density of the prey population, that is bxy. Without prey, predators die at the rate dy. These assumptions give rise to two ordinary differential equations:

$$\dot{x} = rx - axy,$$
$$\dot{y} = bxy - dy. \tag{1.1}$$

Here $\dot{x}$ and $\dot{y}$ denote the derivative with respect to time that is the rate of change over time; x and y are variables that is quantities which change over time, whereas r, a, b, and d are parameters. The equilibrium of the equation is given by $\dot{x} = 0$ and $\dot{y} = 0$, which leads to $x^* = d/b$ and $y^* = r/a$ for the equilibrium values of prey and predators. Let us consider the ratio of predators over prey, $\rho = y^*/x^*$. We have $\rho = (rb)/(ad)$. Fishing effectively reduces the growth rate of prey, and so ρ declines during periods of heavy fishing. Conversely, ρ increases when fishing declines. D'Ancona had his answer, and the world had mathematical ecology.

Volterra studied his model in great depth. His work was greeted with some scepticism by biologists, but was extended by three famous Russian mathematicians: Gause,

Kostizin and Kolmogoroff. It also turned out that the American maverick Alfred J. Lotka had looked at similar equations. Therefore these equations are now known as Lotka–Volterra equations. They are all abundant in problems of modern ecology and beyond. We will see that many equations that we encounter in this book will be related to Lotka–Volterra equations. In our context, the virus will be prey and the immune system predator.

Lotka–Volterra equations display rich dynamics. The basic system, given by eqn (1.1), is characterized by undamped oscillations around a 'neutrally stable' equilibrium. Prey and predators will forever remain in periodic oscillations, their trajectories will not converge toward the equilibrium; only their time average will approximate the equilibrium value. Such a situation is *structurally unstable*: small modifications of eqn (1.1) will make the equilibrium either stable or unstable.

A stable equilibrium implies that over time the trajectories of the dynamical system will converge toward this equilibrium. The approach toward equilibrium can either be direct or in damped oscillations. An equilibrium is unstable if the trajectories lead away from it. If the system starts in an unstable equilibrium it remains there, but any perturbation would lead the system away from this equilibrium.

Lotka–Volterra equations for more than just two species can have several equilibria at once. Some of them may be stable some of them may be unstable. They can also have 'limit cycles': periodic trajectories that close unto themselves. Like equilibrium points, limit cycles can be stable or unstable. A stable limit cycle attracts nearby trajectories, an unstable repels them. In contrast to the simple neutral oscillations, limit cycles are structurally stable. Small modifications of the equations do not destroy them.

Lotka–Volterra equations also admit 'chaotic attractors'. These are aperiodic trajectories which never close unto themselves. The system shows irregular oscillations, which are genuinely unpredictable in the long-term. 'Chaos' as a feature of simple deterministic mathematical equations was an important discovery that led to a frameshift in our understanding of the world.

The application of mathematical models to disease also has a prominent forerunner: in 1760, the celebrated polymath Daniel Bernoulli developed a mathematical method to study the effectiveness of variolation techniques against smallpox. His aim was to influence public health policy. In 1840, William Farr fitted curves to data of smallpox epidemics in England and Wales. At the beginning of this century Hamer and Ross formulated mathematical equations to describe the spread of infectious agents within populations. Hamer introduced the notion of the 'mass action principle', which states that the net rate of spread of infection is proportional to the product of the densities of infected and susceptible individuals. Ronald Ross, describing the spread of malaria, was the first to use a continuous time model. In 1927, Kermack and McKendrick laid the foundations for a theoretical framework of epidemiology.

Again, in the pages of this book, we will find many models similar to the basic models of epidemiology; susceptible and infected individuals will be replaced by uninfected and infected target cells of the virus. In effect, we describe micro-epidemiology, the spread of an infectious agent from cell to cell within one patient.

In the meanwhile mathematical models have been developed for virtually all areas in biology, and many interesting contributions have been made. Nevertheless, mathematical biology still has the air of a fascinating new field, which is full of open questions and unexplored areas. The main reason for this is the explosive increase of empirical data in biology always posing new questions that can benefit from mathematical models.

'Only theory can tell us what to measure and how to interpret it' is one of the many famous one-liners attributed to Albert Einstein. The meaning is clear. Understanding of the world in scientific terms is to build a model, to reduce apparent complexity to a set of simple rules. These rules constitute a theory. A theory may be verbal or in terms of mathematical equations, but a verbal theory is always incomplete. A mathematical theory provides a logical link between assumption and conclusion. Thus, ultimately the language of all natural sciences is mathematics. A verbal theory can be conveniently vague about its details and hide important assumptions. A mathematical theory is more transparent. It contains a clear list of assumptions, which are its ingredients generated by observation.

Mathematical theories and experimental observations go hand in hand. Theory cannot exist without experiment. Experiment is the basis of the scientific method. Theory is relevant once observations have been made. Ultimately, the role of theory is to reduce apparent complexity. Understanding a biological system means to be able to describe it in simple, logical terms. Listing all the factors that contribute to the behaviour of a biological system is an important first step, but scientific progress is to understand which of these factors are relevant. This is what theory is about: selecting a (small) number of assumptions, expressing them in the language of mathematics and following the model rigorously to its conclusion. Then testing these conclusions against empirical facts. Then, most often, refining the assumptions and going through the loop again. And again.

1.4 Further reading

Viruses by Arnold Levine (1992) is an excellent introduction for a general audience. Comprehensive treatments of virology are offered by Dimmock and Primrose (1994) and by Fields and Knipe (1996). See also Diekmann and Heesterbeek (2000). There are many excellent textbooks on immunology; we mention Klein and Horejsi (1997), Roitt *et al.* (1998), and Janeway *et al.* (1999). Mathematical biology without equations is explained by Karl Sigmund in *Games of life* (1995). Some fundamental books of mathematical biology are May (1973), Maynard Smith (1982), Kimura (1983), Edelstein-Keshet (1988), Feldman and Karlin (1989), Murray (1993), Maynard Smith (1998) and Hofbauer and Sigmund (1998). Anderson and May (1991) offers a comprehensive treatment of mathematical epidemiology. Levin *et al.* (1997, 1999) are excellent reviews of population biology and infectious disease.

2

HIV

2.1 Discovery

In 1981 the Centers for Disease Control in Atlanta noticed an increase in rare diseases that are associated with suppression of the immune system. This observation was made in groups of homosexual men in Los Angeles, San Francisco and New York. Patients died from infectious diseases that were normally controlled by the immune system. There was also an increase in a rare cancer, called Kaposi's sarcoma. The number of patients who succumbed to this new immunodeficiency disease increased dramatically. It became clear that a completely new disease with an unknown mechanism had suddenly appeared.

The new disease, which was characterized by a severe impairment of the immune system and related opportunistic infections, was called acquired immune deficiency syndrome (AIDS). At the end of 1982 there were 800 AIDS cases in the United States. The disease had spread to over 30 states and was clearly not restricted to homosexual men, but was found among hemophiliacs, who receive frequent injection of blood derived products, intravenous drug abusers and their heterosexual partners. The disease had also been transmitted from mothers to newborn children. It became clear that the disease was infectious and the route of transmission was via contact with genital secretions or blood. Scientists began a feverish search for the infectious agent of AIDS. Was it a new virus, or a variant of an already known virus? Some people even held the view that AIDS was not caused by a single infectious agent, but was the combination of different diseases.

The first breakthrough came in 1983 when researchers at the Institute Pasteur in Paris, led by Francoise Barre-Sinoussi and Luc Montagnier, demonstrated the presence of a retrovirus in tissue taken from enlarged lymph nodes of an AIDS patient. A retrovirus uses an enzyme called reverse transcriptase to copy its genome from RNA into DNA. At that time only two human retroviruses were known: the human T-cell leukemia viruses (HTLV) I and II.

The world's leading expert on human retroviruses was Robert Gallo at the National Institute of Health in Bethesda. His group had discovered HTLV-I and -II, and he was now on the trace of the AIDS virus. Soon after the French group, Gallo's team also isolated a retrovirus from AIDS patients.

Initially, both Gallo and Montagnier believed that the AIDS virus was closely related to HTLV-1, but while Gallo continued to adhere to this view, Montagnier soon obtained evidence that the AIDS virus belonged to a completely different class of retroviruses, so called lentiviruses. Montagnier's view should turn out to be correct.

Gallo's group, however, succeeded to multiply the AIDS virus in continuous cell cultures which allowed to generate sufficient amounts of virus to develop a specific test for the virus. This was an important achievement, because with such a test in hand, doctors and scientists were no longer faced with an invisible opponent. It was now possible to test whether a given patient was infected by the new virus, and whether blood products to be used for transfusion were contaminated with virus.

Once the causative agent of the AIDS epidemic was identified, optimism for finding a treatment arose. Scientists all over the world began to reveal important features of the new virus at a rapid pace, and at a press conference in 1984, Robert Gallo predicted that an AIDS vaccine would be available within two years.

Gallo had named the AIDS virus HTLV-III, whereas the French scientists coined the name LAV, for lymphadenopathy-associated virus. A British group led by Robin Weiss showed that HTLV-III and LAV were the same type of virus. They also discovered that it used the CD4 cell receptor to make its way into the cells of the immune system. It was agreed that the new virus should be called human immunodeficiency virus (HIV).

2.2   Some basic facts about HIV

HIV is a retrovirus, which means that its genome is in form of RNA and is translated into DNA during its life-cycle. The translation of the viral RNA into DNA is conducted by a specific viral enzyme called reverse transcriptase.

The central dogma of molecular biology formulated by Francis Crick in the 1960s stated that information in biology always flows from DNA to RNA to protein. The discovery of reverse transcription, which passes information from RNA back to DNA, was a violation of one step of the central dogma and won a Nobel prize for its discoverers Howard Temin and David Baltimore. Importantly, it also presented a new way of viral replication and explained how certain viruses with RNA genomes could insert a copy of their genetic information in form of DNA into the genome of the host cell.

There are three types of retroviruses:

(i) Oncoviruses cause cancers. They insert their genome into the host cell and manipulate it to divide excessively, which can lead to malignant growth of the host cell tissue. HTLV-1 and HTLV-2 are human oncoviruses.

(ii) Lentiviruses are characterized by a relatively slow development of symptoms and disease. HIV is a lentivirus. So are simian immunodeficiency viruses (SIV), feline immunodeficiency viruses (FIV), visna medi virus of sheep, the equine infectious anaemia virus and others.

(iii) Spumaviruses are the least well characterized retroviruses. They have been isolated from humans and other mammal species, but do not seem to be associated with any disease.

The total genome size of HIV is about 10 000 bases and contains nine genes in different reading frames. There are three structural genes called *gag*, *pol* and *env*. The *gag* gene encodes for a protein which is cleaved by the HIV protease into four subunits; these subunits form the coat that encloses the genetic material of the virus. The *pol* gene encodes for the reverse transcriptase, the protease, and the integrase; the reverse transcriptase copies the viral RNA into DNA, the protease cleaves the *gag* gene product and the integrase helps to insert the viral DNA into the genome of the host cell. The *env* gene encodes two viral envelope proteins called gp-41 and gp-120. Gp-41 is embedded in the membrane of the HIV particle, whereas gp-120 sits on the outside of the viral membrane attached to gp-41. Gp-120 is essential for the virus to enter its host cell.

In addition to these structural genes, there are six regulatory genes, which encode for proteins that are involved in regulating viral replication inside the cell and manipulating the machinery of the cell.

The main target cell of HIV is the CD4 positive T helper cell. The viral envelope protein gp-120 binds to the CD4-cell receptor. Subsequently, viral and cellular membranes fuse and the core of the virus including the genetic material enters the cell. In every virus particle, there are two copies of the RNA genome. (Diploidy is a feature of all retroviruses.) Once inside the cell, the reverse transcriptase, which is present in multiple copies in the core of the virus particle, starts to copy the viral RNA into DNA. Interestingly, it uses both RNA molecules for this process and jumps frequently from one molecule to the other thereby producing a DNA molecule which is a recombination product of the two RNA molecules. The double stranded DNA is then inserted into the genome of the host cell.

Subsequently, the genetic machinery of the host cell is manipulated to make RNA copies of the viral genome. Some of these RNA molecules are used as templates for the production of viral proteins, while the others are used as new viral genomes. The viral envelope proteins, gp-41 and gp-120, are inserted into the cell membrane. The new viral particles are assembled on the inside of the cell membrane; they leave the host cell by budding from the membrane taking the host cell membrane as new viral envelope. The host cell usually dies when thousands of virus particles bud from its surface.

It is an unfortunate coincidence that HIV is adapted to infect CD4 positive T helper cells which are meant to take a key position in fighting infectious agents. A healthy human adult has about 1000 CD4 cells per μl of blood, but in an HIV infected patient the abundance of CD4 cells can decline to very low numbers. A CD4 cell count of less than 200 is currently used as definition for AIDS regardless of the presence of opportunistic infections.

The pattern of disease progression in HIV infection is subdivided into three phases: (i) The first few weeks after inoculation with virus mark the primary phase of the infection. Patients usually develop high virus loads (which is the abundance of virus in the blood) and show some symptoms of a virus infection often described as being similar to flu. In this phase, the virus is distributed to many different organs of the body. CD4 cells fall transiently and then return to almost normal levels. At the end of the primary phase, virus load falls again and the clinical symptoms disappear. (ii) Patients enter the second phase of HIV disease which is largely asymptomatic, but the virus continues to replicate and CD4 cells fall relentlessly. In some patients CD4 cells decline rapidly in others more slowly. There is great heterogeneity in the length of the asymptomatic phase ranging from a few months to many years. The average length of the asymptomatic period is about 10 years. (iii) The final phase of the disease is characterized by the development of AIDS. CD4 cells have fallen to below 200 per μl, and opportunistic infections begin to appear. The impaired immune system of the patient can no longer fight off infections that would normally pose no problem. Patients die from opportunistic infections.

Despite the apparently inevitable progression to disease and death, many HIV infected patients have strong immune responses against the virus. Antibodies against the envelope protein gp-120 can prevent virus particles from infecting cells. Killer cell responses against peptides of various viral proteins can eliminate virus infected cells. Killer cell responses are usually found in asymptomatic patients, but seem to be weak or absent in AIDS patients. While it still unresolved to what extent immune responses slow down disease progression in HIV, it is obvious that they are not sufficiently strong to eliminate the virus. A main part of the book will be concerned with the mechanisms how HIV escapes from immune responses.

There is a good correlation between virus load and rate of disease progression: the survival time of HIV infected patients is correlated to the virus load during the first year of infection. A high virus load implies rapid disease progression, a low virus load slow disease progression. It is now widely accepted that virus load is directly linked to disease in HIV infection.

When an HIV particle attaches to a target cell, its envelope protein gp-120 binds to the CD4 cell receptor, but Robin Weiss' first studies made it clear that there was a second receptor involved on the side of the target cell. The CD4 receptor was necessary but not sufficient to permit virus entry. The secondary receptor(s) eluded discovery for many years until a decisive experiment performed in Robert Gallo's lab demonstrated that certain chemicals, called chemokines, can prevent virus particles from entering cells. It turned out that the elusive secondary receptors were proteins that bind to chemokines, so-called 'chemokine receptors'. Chemokines are signal substances between cells of the immune system. There are several different chemokine receptors on CD4 T cells and on macrophages. There are different variants of the HIV virus that use different secondary receptors. This poses interesting questions for the evolution of cell tropism during HIV infection. (The 'cell tropism' of a virus is defined by the types of cells that can be infected by this virus.) Discovery of the secondary receptors may also reveal a new method for anti-HIV therapy: chemokines may be used therapeutically to prevent the virus from entering cells.

Soon afterwards, Stephen O'Brien and colleagues at the National Cancer Institute discovered genetic differences in humans in the chemokine receptor and thereby identified individuals that are virtually resistant to infection by the HIV virus. These individuals have a deletion in one of the chemokine receptors. In the Caucasian population about 15% of individuals may be resistant to HIV, but in the African population almost 0%.

2.3   Treatment

The recent years have seen enormous progress in the design of specific drugs that inhibit HIV replication within patients. The first successful anti-HIV drug was zidovudine (or AZT). It was originally designed as anti-cancer drug aimed at preventing rapid cell division, but it turned out to have a specific effect on the HIV encoded reverse transcriptase enzyme. It inhibits this enzyme and thereby prevents the viral RNA from being copied into DNA, an essential part of the HIV life-cycle.

In 1989, clinical trials with zidovudine in HIV patients were stopped prematurely, when it became clear that the drug had a beneficial effect. Unfortunately, the effect of zidovudine is only short lived; AIDS patients have an increased life expectancy of about six months. The main reason of the limited success of zidovudine seems to be viral resistance. HIV can mutate to become resistant against the drug. One or a few point mutations in the reverse transcriptase gene can render the enzyme insensitive to zidovudine. The altered enzyme no longer binds to the drug.

In the meanwhile several new reverse transcriptase inhibitors have been designed, some are more potent than zidovudine. Reverse transcriptase inhibitors fall into two categories: (i) nucleoside analogues (like zidovudine) act as chain terminators; they are incorporated into the growing DNA molecule and prevent its extension; (ii) non-nucleoside analogues bind to another part of the reverse transcriptase enzyme and stop it from functioning.

Furthermore, another virus enzyme, the protease has been targeted. Protease inhibitors were designed on the computer to bind to the HIV protease and prevent it from cleaving the gag precursor protein, which results in the production of non-infectious virus particles.

All drugs if used individually have the same problem: the virus readily develops resistance. But success came when several drugs were used simultaneously. A combination of zidovudine and another nucleoside analogue, called lamivudine, led to a long-term 10-fold suppression of virus load in many patients. The virus apparently could not come up with a mutant completely resistant to both drugs.

This was followed by triple-drug therapy: a protease inhibitor is combined with two reverse transcriptase inhibitors. In about 1995, it was possible by means of 'combination therapy' to reduce virus load in patients below the limit of detection (about 50 copies of viral RNA per μl blood) and keep it undetectable for many years. HIV patients, for the first time, were given hope to live for many years.

2.4   Origins of HIV

The human immunodeficiency virus comes in two types: HIV-1 and HIV-2 share about 45% sequence homology in their genome. HIV-2 occurs mostly in West Africa, whereas HIV-1 has spread all over the world. The closest relatives of HIV are the SIV infecting a large variety of non-human primates. Apparently all primate species that have been studied harbour an SIV virus. Remarkably, all SIV viruses seem to be apathogenic in their natural hosts.

The closest relative of HIV-2 is SIV from sooty mangabeys that live in West Africa; they have almost identical sequence. The equivalent monkey virus for HIV-1 was not known for many years. Only in 1998, Beatrice Hahn and George Shaw discovered a virus nearly identical to HIV-1 in chimpanzees.

SIV viruses have been with their hosts for millions of years. It is likely that humans were always exposed to these viruses and occasionally got infected, but the viruses did not specifically adapt to humans and did not cause an epidemic. Maybe after a few transmissions the virus disappeared again from the human population.

In this century, however, the situation in Africa may have changed. A dramatic increase in the population size and flexibility may have provided sufficient ground for the virus to grow. It got transmitted to more humans and had the possibility to adapt to grow well in human cells. There are also speculations that mass vaccination programs in Africa distributed the virus in the initial phase of the epidemic. From Africa, the HIV virus got distributed to all other parts of the world. The pandemic had started.

2.5 Further reading

Excellent books on HIV and retroviruses include Coffin *et al.* (1997), Goudsmit (1997), and Levy (1998). Also see Diamond (1997). Some important, very early papers on HIV are Barre-Sinoussi *et al.* (1983), Gallo *et al.* (1983), Gelman *et al.* (1983), Dalgleish *et al.* (1984), Hahn *et al.* (1984), Levy *et al.* (1984), Shaw *et al.* (1984), Chiu *et al.* (1985), Fisher *et al.* (1985), Robert-Guroff *et al.* (1985), Shaw *et al.* (1985), Wain-Hobson *et al.* (1985), Weiss *et al.* (1985*a*), Wong-Staal and Gallo (1985), Wong-Staal *et al.* (1985), Clavel *et al.* (1986, 1986*a*), Feinberg *et al.* (1986), Haase (1986), Hahn *et al.* (1986), Koenig *et al.* (1986), Lasky *et al.* (1986), Lifson *et al.* (1986), Lifson *et al.* (1986*a,b*), Sattentau *et al.* (1986), Walker *et al.* (1986), Weiss *et al.* (1985*a,b*, 1986), and Zagury *et al.* (1986). Incubation period of HIV infection was measured by Biggar (1990). For resistance to AZT refer to Larder and Kemp (1989). Mathematical modelling of AZT resistance was done by McLean and Nowak (1992*a*). Samson *et al.* (1996) and O'Brien and Dean (1997) describe resistant genes in HIV infection. For work on HIV coreceptors see Cocchi *et al.* (1995), Dragic *et al.* (1996), Connor *et al.* (1997), Dittmar *et al.* (1997), Gallo and Lusso (1997), Rowland-Jones and Tan (1997). Matloubian *et al.* (1994) describe CD4 help during chronic viral infection.

3

THE BASIC MODEL OF VIRUS DYNAMICS

Traditional epidemiology describes the spread of infectious agents within populations of individuals. Here we turn to the population dynamics of infection at a different level, which happens within the body of an infected individual; we describe how viruses spread from cell to cell. This micro-epidemiology opens up a completely new perspective for understanding infectious diseases. The aim is to reveal the basic laws that govern the spread of infectious agents within individuals, their interaction with the immune system and their response to treatment.

In this chapter, we formulate the simplest model of virus dynamics and demography. This serves as a point of departure for most of our investigations. The basic model has three variables: uninfected cells, infected cells and free viruses. A system of three differential equations describes how these quantities change over time. We assume that uninfected cells encounter free virus and turn into infected cells. In this simplest model, the rate of production of new infected cells is proportional to the product of the density of uninfected cells times the density of free virions. Free virions are produced by infected cells. Uninfected cells, infected cells and free virions die at certain rates. In addition, we assume that uninfected cells are constantly replaced by the system. These assumptions define the basic model of virus dynamics.

Suppose the system is initially in an uninfected state. Uninfected cells are at a certain equilibrium value. Let us add a small amount of virus particles (or infected cells). The invading virions will manage to infect a number of cells, which will produce new virions, which will infect new cells. A chain reaction has started. The chain reaction can go two ways: either it dies out or it leads to a massive explosion of free virions and infected cells. Whether or not the chain reaction will take off, depends on a quantity called the basic reproductive ratio of the infection, which is denoted by R_0.

R_0 is essentially the ecologists' 'basic reproductive rate', which is fundamental to any discussion of the demography of populations of living things, be they humans, frogs, oaks, helminths, protozoa, or whatever (actually, of course, R_0 is a dimensionless ratio, not a rate, but this point has eluded ecologists). For a viral infection, R_0 is the average number of infected cells that derive from any one infected cell in the beginning of the infection. If every infected cell produces on average less than one newly infected cell, that is $R_0 < 1$, then the infection will not take off. The initial inoculum of virus will disappear. If every infected cell does produce more than one newly infected cell, that is $R_0 > 1$, then the infection will take off.

In this case, virions and infected cells will initially grow exponentially, while uninfected cells stay roughly constant. After some time the abundance of uninfected cells will decline and the exponential growth of virus will slow down. The abundance of virus will reach a maximum and subsequently decline. Simultaneously, the abundance of uninfected cells will pass through a minimum and then increase again. The system will converge, in damped oscillation, to an equilibrium point.

At equilibrium, the virus is controlled by a limited supply of uninfected cells. Every infected cell will, on average, give rise to one newly infected cell. The equilibrium abundance of uninfected cells is therefore given by the number of uninfected cells prior to infection divided by R_0. Thus if R_0 is much bigger than 1, the abundance of uninfected cells will be greatly reduced.

Another interesting feature of the model is that a cytopathic virus, which rapidly kills infected cells, will only achieve a small equilibrium abundance of infected cells, no matter how many virions are being produced from any one cell. In contrast, a non-cytopathic virus will lead to an equilibrium abundance of infected cells close to the total abundance of cells prior to infection. In both cases, the rate at which virions can infect target cells has only a small effect on the equilibrium abundance of infected cells. Many biologists have found these facts to run counter to their intuition: they confuse the rate at which water flows into and out of the bath (birth and death rate of infected cells) with the level of water in the bath (equilibrium number of infected cells).

The basic model can also be used to describe the primary phase of HIV or SIV infection. During the first weeks of infection there is a peak in virus load with a subsequent decline to a relatively stable steady-state. The model helps to estimate rate constants of the initial virus expansion and decline. It is also possible to get an idea of the basic reproductive ratio, R_0, during this first phase of infection. Furthermore, extensive data from SIV infection show a strong correlation between the virus growth rate during the first week of infection and the steady-state virus load after the initial peak, which in turn affects the rate of progression to disease and death.

The remainder of this chapter is divided in four parts, which give flesh to the skeleton we have just sketched. First, we will specify the assumptions and equations of the basic model. Second, we will describe the dynamics. Third, we will analyse the equilibrium properties. Fourth, we will apply the model to the dynamics of the primary phase of HIV and SIV infection.

3.1   The model

The basic model of virus dynamics (Fig. 3.1) has three variables: the population sizes of uninfected cells, x; infected cells, y; and free virus particles, v. These quantities can either denote the total abundance in a host, or the abundance in a given volume of blood or tissue.

Free virus particles infect uninfected cells at a rate proportional to the product of their abundances, $\beta x v$. The rate constant, β, describes the efficacy of this process, including the rate at which virus particles find uninfected cells, the rate of virus entry, and the rate and probability of successful infection. Infected cells produce free virus at a rate proportional to their abundance, ky. Infected cells die at a rate ay, and free virus particles are removed from the system at rate uv. Therefore, the average life-time of an infected

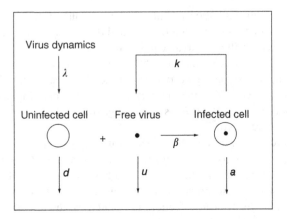

Fig. 3.1 Schematic illustration of the basic model of virus dynamics. Uninfected cells 'react' with free virus to give rise to infected cells; the rate constant is β. Infected cells produce free virions at a rate k. Uninfected cells, free virus and infected cells die at the rates d, u and a, respectively. Uninfected cells are replenished at a constant rate λ.

cell is $1/a$, whereas the average life-time of a free virus particle is $1/u$. The total amount of virus particles produced from one infected cell, the 'burst size', is k/a.

In addition to the dynamics describing virus infection, we have to specify the dynamics of the uninfected cell population. The simplest assumption is that uninfected cells are produced at a constant rate, λ, and die at a rate dx. The average life-time of an uninfected cell is therefore $1/d$. In the absence of an infection, the population dynamics of host cells is given by $\dot{x} = \lambda - dx$. This is a simple linear differential equation. Without virus, the abundance of uninfected cells converges to the equilibrium value λ/d.

Combining the dynamics of virus infection and host cells, we now obtain the basic model of virus dynamics:

$$\begin{aligned}
\dot{x} &= \lambda - dx - \beta xv, \\
\dot{y} &= \beta xv - ay, \\
\dot{v} &= ky - uv.
\end{aligned} \tag{3.1}$$

This is a system of nonlinear differential equations. An analytic solution of the time development of the variables is not possible, but we can derive various approximations and thereby obtain a complete understanding of the system.

3.2 Dynamics

Before infection, we have $y = 0$, $v = 0$, and uninfected cells are at equilibrium $x = \lambda/d$. Denote by $t = 0$ the time when infection occurs. Suppose infection occurs with a certain amount of virus particles, v_0. Thus the initial conditions are $x_0 = \lambda/d$, $y_0 = 0$ and v_0.

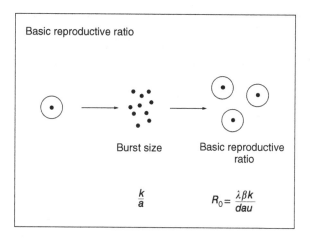

Basic reproductive ratio

Burst size   Basic reproductive
                           ratio

$$\frac{k}{a} \qquad\qquad R_0 = \frac{\lambda\beta k}{dau}$$

Fig. 3.2 The burst size is the total number of virions produced from any one infected cell. The basic reproductive ratio is the total number of newly infected cells that arise from any one infected cell in the beginning of the infection (when the abundance of uninfected cells is still at its pre-infection level).

Whether or not the virus can grow and establish an infection depends on a condition very similar to the spread of an infectious disease in a population of host individuals. The crucial quantity is the basic reproductive ratio, R_0, which is defined as the number of newly infected cells that arise from any one infected cell when almost all cells are uninfected (Fig. 3.2). The rate at which one infected cell gives rise to new infected cells is given by $\beta k x / u$. If all cells are uninfected then $x = \lambda / d$. Since the life-time of an uninfected cell is $1/a$, we obtain

$$R_0 = \frac{\beta \lambda k}{adu}. \tag{3.2}$$

If $R_0 < 1$ then the virus will not spread, since every infected cell will on average produce less than one other infected cell. The chain reaction is sub-critical. If we start with N infected cells then, on average, we expect roughly $\ln N / \ln(1/R_0)$ rounds of replication before the virus population dies out.

If on the other hand $R_0 > 1$, then every infected cell will on average produce more than one newly infected cell. The chain reaction will generate an explosive multiplication of virus

$$v(t) = v_0 e^{r_0 t}, \tag{3.3}$$

where the exponential growth rate of the virus population, r_0, is given by the larger root of the equation $r_0^2 + (a+u)r_0 + au(1 - R_0) = 0$. If $u \gg a + r_0$, we find the approximation $r_0 = a(R_0 - 1)$. This can also be written as $r_0 = \beta' x_0 - a$, where $\beta' = \beta k / u$; in other words, $r_0 = a(R_0 - 1)$, which says that each infected cell is producing R_0 newly infected cells before dying after an average life-time of duration $1/a$; this corresponds to a net rate of $a(R_0 - 1)$. Virus growth will not continue indefinitely, because the supply of uninfected cells is limited. There will be a peak in virus load and subsequently damped oscillations to an equilibrium (Fig. 3.3).

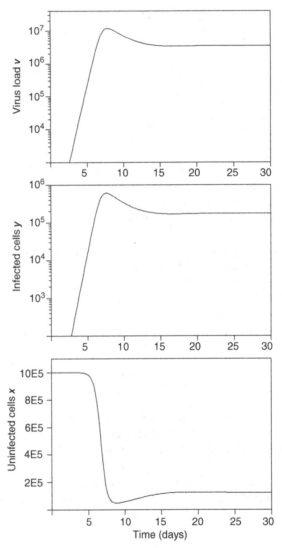

Fig. 3.3 The basic model of virus dynamics shows an exponential growth phase of free virus and infected cells followed by a peak and subsequent decline to an equilibrium. Uninfected cells stay initially constant, then decline sharply to a nadir and subsequently recover to their equilibrium value. These dynamics reflect the primary phase of HIV or SIV infection. Parameter values of the simulation: $\lambda = 10^5$, $d = 0.1$, $a = 0.5$, $\beta = 2 \times 10^{-7}$, $k = 100$, $u = 5$.

3.3 Equilibrium

The equilibrium abundance of uninfected cells, infected cells and free virus is given by

$$x^* = \frac{au}{\beta k} = \frac{x_0}{R_0},$$

$$y^* = (R_0 - 1)\frac{du}{\beta k}, \qquad (3.4)$$

$$v^* = (R_0 - 1)\frac{d}{\beta}.$$

At equilibrium, any one infected cell will on average give rise to one newly infected cell. The fraction of free virus particles that manage to infect new cells is thus given by the reciprocal of the burst size, a/k. The probability that a cell (born uninfected) remains uninfected during its life-time is $1/R_0$. The equilibrium ratio of uninfected cells before and after infection is $x_0/x^* = R_0$.

If the virus has a basic reproductive ratio much larger than one, then x^* will be greatly reduced compared to x_0, which means that during infection the equilibrium abundance of uninfected cells is much smaller than before infection. In other words, the above simple model cannot explain a situation where during a persistent virus infection almost all infectable cells remain uninfected ($x^* \approx x_0$), except in the case when R_0 is only slightly bigger than unity (which is *a priori* unlikely in general).

Furthermore, if $R_0 \gg 1$, then the equilibrium abundance of infected cells and free virus is approximately given by $y^* \approx \lambda/a$ and $v^* \approx (\lambda k)/(au)$. Interestingly, both quantities do not depend on the infection parameter β. The reason is that a highly infectious virus (large β) will rapidly infect uninfected cells, but at equilibrium there will only be few uninfected cells in the system. A less infectious virus (smaller β) will take longer to infect uninfected cells, but the equilibrium abundance of uninfected cells is higher. For both viruses the product βx will be the same at equilibrium, resulting in a constant rate of production of new infected cells, and therefore in similar equilibrium abundances of infected cells and free virus.

For a highly cytopathic virus (a much larger than d), the equilibrium abundance of infected cells will be small compared to the abundance of cells prior to infection. In fact, the larger a, the smaller the abundance both of infected cells and of free virus. This result, although mathematically trivial, is an important one.

For a non-cytopathic virus ($a \approx d$), the equilibrium abundance of infected cells will be roughly equivalent to the total abundance of susceptible cells prior to infection.

3.4 The primary phase of HIV and SIV infection

In primary HIV infection, a peak of plasma virus load is typically followed by a decrease in levels of circulating virus of up to 2–3 orders of magnitude over the succeeding few weeks. While cellular and antibody responses have been suggested as the mechanisms responsible for this reduction in virus load, it is interesting to note that the simplest model of virus dynamics does not require the involvement of specific immune responses to reflect the observed dynamics. Virus load may decline simply because of a shortage of susceptible cells.

Several studies have clearly demonstrated that in most HIV infected subjects, following resolution of acute primary infection, plasma levels of virus equilibrate and remain relatively stable over the short term. The level at which plasma viremia stabilises in the post-acute infection period is set by the balance between the ongoing viral replication and clearance, and has been shown to be an important prognostic determinant. Higher levels of plasma viremia at 6–12 months following primary infection are associated with a significantly increased relative risk for more rapid disease progression.

Studies in SIV infected macaques have also documented extensive and continuous viral replication throughout the course of infection, as well as correlations between different patterns of viral replication and clinical course. In contrast to the study of acute HIV infection in human subjects, in experimental SIV infection it is possible to control the size and exact timing of the initial inoculum and to obtain several samples from the primary phase of the infection.

In one study, 12 pigtail macaques (*M. nemestrina*) were inoculated with the same amount of a specified strain of SIV (Fig. 3.4). There was extensive variation in virus growth rate during the first week of infection. At day 4, 4 of 12 animals had detectable plasma virus load (that is more than 400 copies of viral RNA per ml). At day 7, all animals had detectable virus load, ranging from 58 000 to 1.3 million copies per ml. The average doubling time of virus load was about 8 h. Peak levels of virus load in individual animals were reached between days 10 and 21 after inoculation and ranged from 800 000 to 7 million copies per ml. Subsequently virus load declined to lower levels. The 'set-point' equilibrium virus load was estimated by taking the average of virus load measurements at days 36, 38 and 42; it ranged from 3900 to 2.6 million in different animals.

Interestingly the data reveal a strong correlation between virus growth in the first week of infection and the equilibrium virus load. This is of particular importance since earlier studies (both for HIV and SIV) have shown that virus load early in infection (after the initial peak) determines the rate of progression to disease and death. Individuals with low virus load live longer. Thus it seems that a prediction for the rate of disease progression can already be made from data in the first week of infection.

What is the implication of the observed correlation? Under which conditions does the basic model of virus dynamics predict a correlation between initial growth and equilibrium virus load?

If $R_0 \gg 1$ we have approximately

$$v(t) = v_0 \exp\left[\frac{\beta \lambda k}{du} t\right] \tag{3.5}$$

during the exponential growth phase and

$$v^* = \frac{\lambda k}{au} \tag{3.6}$$

at equilibrium. Let us analyse the correlation between $v(t)$ and v^* if animals differ in individual parameters of virus dynamics.

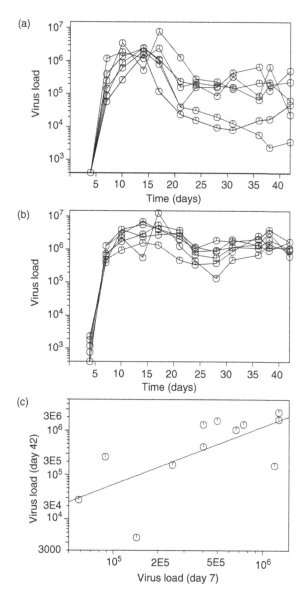

Fig. 3.4 Experimental data of SIV primary infection. Twelve macaques were infected with the same amount of virus. (a) Plasma virus of six animals with virus load set-point below 10^6 (average virus load at day 36, 38 and 42). (b) Plasma virus of six animals with a set-point above 10^6. (c) Highly significant correlation between virus load at days 7 and 42 in 12 animals. Data are taken from Lifson *et al.* (1997) and Nowak *et al.* (1997).

Suppose animals differ mostly in the rate of supply of uninfected cells, λ, while all other parameters are roughly constant. In this case, the model predicts a linear relation between equilibrium virus load, v^*, and the logarithm of virus load, $v(t)$, at time t during the exponential growth phase:

$$v^* = \frac{d}{\beta a t}[\ln v(t) - \ln v_0]. \tag{3.7}$$

More precisely, this correlation holds if variation among animals is confined to the parameters λ, k, or u, while β, a, and d are roughly constant.

Variation in β, the rate of infection of uninfected cells, or in d, the death rate of uninfected cells, interestingly only causes variation in $v(t)$, but does not affect the equilibrium virus load (as long as $R_0 \gg 1$). Variation in a, on the other hand, only affects the equilibrium virus load, but not the exponential growth phase. Thus variation in any one of the parameters β, d, or a does not lead to a correlation between $v(t)$ and v^*.

While the model can explain the observed correlation if animals differ in λ, k or u, it does predict more variation to occur during the exponential growth phase then at equilibrium, because differences in any of the parameters enter exponentially into $v(t)$ but only linearly into v^*. The data, however, suggest the opposite. There is a wider range of virus loads at steady-state than at day 7. The likely explanation is that the additional variation at equilibrium is caused by the development of specific immune responses.

Thus the experimental observations in conjunction with the model suggest the following scenario. There are important differences in animals in parameters which affect the initial rate of virus growth during the first week of infection. These parameters are still important in determining virus load much later (days 36–42), when an equilibrium has been reached. Other factors (such as specific immune responses) which do not affect the initial virus growth, because they develop later, introduce additional variation into the virus load at steady-state, but their contribution does not destroy the correlation. If virus load at steady-state depended only on such factors, then we would not expect the observed correlation.

In a subsequent study of SIV infection, it was possible to show a correlation between the rate at which the virus grows *in vitro*, in cells taken from animals prior to infection, and the *in vivo* growth rate of the virus during the first week of infection. This suggests that the abundance of target cells or their permissiveness for virus growth are key parameters that determine both the initial and long-term pattern of virus growth. In terms of our model, the abundance of target cells is determined by the parameter, λ, while their permissiveness is determined by the rate of virion production, k.

3.4.1   *Estimating R_0*

Using the basic model of virus dynamics it is possible to estimate the basic reproductive ratio, R_0, for SIV infection. The relation between R_0 and the initial exponential growth rate, r_0, is given by $R_0 = 1 + r_0(r_0 + a + u)/au$. If $r_0 + a$ is small compared to u, we can write $R_0 = 1 + r_0/a$. Therefore we can directly relate the slope of virus load increase, r_0, during the primary phase to the basic reproductive ratio R_0, provided we have an estimate of a.

The best method to obtain the death rate of infected cell, a, is by analysing data from anti-viral drug treatment as will be discussed in the next chapter. Data from SIV infection suggest $a \approx 0.74$ per day, and r_0 ranging from 0.9 to 2.7 per day. This implies values of R_0 roughly between 2 and 5. Thus in the beginning of the infection, an infected cell generates between 2 and 5 newly infected cells on average.

The standard model of virus dynamics, however, will underestimate the true basic reproductive ratio, if there is a significant time delay between infection of a cell and the time a cell starts to produce new virus particles. In more complicated models of virus dynamics, which include such a time delay, the basic reproductive ratio is generally given by $R_0 = P\beta\lambda k/(adu)$, where P is the probability that an infected cell survives the delay time. If this probability is close to one then there is little difference in R_0 between the standard model and models including delay. What is different, however, is the relation between r_0 and R_0. For a model with a fixed time delay, we find $R_0 = (1 + r/a)e^{r\tau}$. For a model with an exponentially distributed time delay of mean τ we find $R_0 = 1 + (r/a)[1 + (r + a)\tau]$. Both equations are for the limit where u is much larger than a or r. For the second equation we also used the approximation that $1/\tau$ is much larger than d.

Using the same values for r_0 and a as above we obtain estimates for R_0 ranging from 4 to 17 for an exponentially distributed time delay with an average of 1 day. For a fixed time delay of 1 day, the estimate of R_0 is in the range 5–68.

Similar figures were obtained for HIV infection. The main problem here is that the exact time of infection is normally not known. Little et al. (1999) estimated R_0 to range between 7 and 34 including the uncertainty of intracellular delay. Stafford et al. (2000) estimated R_0 to range between 3 and 11, but excluded the complication of intracellular delay, which will significantly increase their maximum estimate, as outlined above.

3.4.2   Vaccination to reduce R_0

Clearly the aim of an anti-HIV vaccine is to reduce the basic reproductive ratio, R_0, to less than one. In this case, the explosive initial increase of virus load could not take place and the person would essentially be protected against infection.

Failing to reach this aim, however, it might still be advantageous to have a vaccine that reduces the initial rate of virus reproduction. If the viral R_0 in a vaccinated person is less than in an unvaccinated person, but still greater than one, then infection will take place, but at a slower rate of viral increase. This could buy important time for the immune system to mount an effective response and subsequently maintain the virus at low levels. Once again we refer to our finding on SIV infection that a slower initial growth rate is correlated with a lower set-point virus load and slower disease progression. A theoretical mechanism for such an association can be that a low initial rate of virus increase does not eliminate an effective anti-HIV CD4 cell response which in turn helps to establish a memory CTL response that can lead to long lasting virus control (Wodarz and Nowak 1999).

The basic message here is that a suboptimal vaccine might not prevent infection, but might help patients to mount effective immune responses and become slow progressors.

3.5 Further reading

The basic model of virus dynamics is discussed in Nowak and Bangham (1996) and Bonhoeffer *et al.* (1997*a*). For the basic reproductive ratio, R_0, in epidemiological theory, we refer to Anderson and May (1979*a,b*, 1991). These publications also describe the basic model of epidemiological dynamics which is, apart from the equation for free virus which hardly ever matters, identical to the basic model of virus dynamics and can be seen as a kind of intellectual ancestor. Lifson *et al.* (1997) and Nowak *et al.* (1997*a,b*) describe data and theory of primary SIV infection. Herz *et al.* (1996) uses a time-delay model for describing viral dynamics. Little *et al.* (1999) and Stafford *et al.* (2000) estimate R_0 for primary HIV infection. Wodarz and Nowak (1999) studied the effect of initial virus growth rate on the establishment of a memory CTL response. Wodarz *et al.* (1999*b*) describe a basic model of HTLV-1 infection.

4

ANTI-VIRAL DRUG THERAPY

In January 1995, George Shaw from the University of Alabama at Birmingham and David Ho from the Aaron Diamond AIDS Research Center in New York and their colleagues published two papers in the journal *Nature*, which hit the headlines more than any other AIDS-related paper in that year. Within a few months almost every HIV virology paper referred to the new findings, and by the end of the year, the papers had established a historical record: compared to all scientific papers ever published they received the largest number of citations in their first year of publication (about 500).

Both Shaw and Ho had each treated about 20 HIV-1 infected patients with new anti-HIV drugs which inhibit an enzyme of the virus called 'protease'. The protease is responsible for cleavage of HIV's inner core protein Gag. A drug that works as a *protease inhibitor* prevents an infected cell from producing the correct form of the Gag protein, and therefore virus particles which leave the cell are malfunctioning; they cannot infect new cells.

Shaw and Ho observed that as a consequence of anti-viral therapy there is a massive reduction in plasma virus load in most patients. Within 2 weeks of treatment the abundance of free virus particles had fallen about 100-fold in most patients. For example, if a patient had around 10^6 virions (= virus particles) in 1 ml of plasma, then after 2 weeks of treatment with a protease inhibitor the viral abundance would have declined to about 10^4 virions per ml. The decay of plasma virus load was to a good approximation exponential. This means if you plot virus load on a logarithmic scale, you obtain a straight line. You can directly read off the slope and obtain the decay rate of plasma virus. The average slope was about 0.4 per day.

Simultaneously with the massive decline in virus load, Ho and Shaw witnessed a rapid increase in CD4 cell numbers in most patients. As we outlined in Chapter 2, CD4 cells are white blood cells which constitute the preferred target for HIV. A healthy person has a CD4 cell count of about 1000 per μl blood; in an AIDS patient this number has dropped to below 200 and can essentially go to zero. The patients treated by Ho and Shaw had initial CD4 cell count between 10 and 500. In many of these patients, CD4 cell counts went up by about 100–400 per μl in less than 2 weeks.

What is important about these findings? Interestingly, the short-term, highly beneficial effect of anti-HIV treatment was nothing new. In fact it had already been known for many years that patients who receive the anti-viral drug zidovudine (also known as AZT) show increasing CD4 cell numbers and declining viral abundance. In earlier experiments, however, viral abundance was only measured indirectly as the plasma concentration of a viral protein, called Gag p24, which is released from ruptured virus particles or from infected cells. In addition, zidovudine is not as potent an inhibitor of virus reproduction as the new protease inhibitors, and only leads to a smaller overall decline of virus load, but the basic observation is essentially the same.

Several factors made Shaw's and Ho's study a new discovery. First, they sampled virus load directly (rather than the indirect and crude measure of p24 protein). Secondly, they obtained sampling points at frequent intervals before and after the start of therapy. Thirdly, and most importantly—in our biased view—they teamed up with mathematical biologists to give the paper a more formal, theoretical twist and to obtain an exact mathematical interpretation of their virus decay data. David Ho had contacted Alan Perelson from the Los Alamos National Laboratories in New Mexico, while George Shaw invited Martin Nowak then at the University of Oxford to join his side.

Shaw sent his data via email. (Theoreticians love experiments, as long as the results arrive by email.) Martin Nowak, working with Sebastian Bonhoeffer, plotted the data logarithmically and read of the slope of virus decline. This can be done with free eye, with a ruler, or with a linear regression program. The result was between 0.3 and 0.6 per day for most patients. There were very few exceptions. Virus load goes down by about 30% per day.

But what does this result tell us? Let us consult the basic model of virus dynamics (see Chapter 3). Infected cells produce free virus. Free virus infects uninfected cells to produce new infected cells. Free virus and infected cells also die at some rate. This can easily be put into equations. An anti-viral drug interferes with this cycle of virus reproduction. To first approximation, it prevents the infection of new cells. What does this do to the equations? It breaks the basic model into two parts. The dynamics of uninfected cells become decoupled from the dynamics of infected cells and free virus. The decay kinetics of free virus and infected cells are now described by linear differential equations.

The basic model of virus dynamics, like almost every other important equation in mathematical biology, is nonlinear. Nonlinear systems are not easy to understand. Linear systems, in contrast, are more straightforward. Every first year undergraduate in mathematics or physics knows this. Linear systems always admit an analytic solution. Nonlinear systems rarely provide theoreticians with the pleasure of analytic solutions. They require more thought and imagination. They can have surprises: chaos and cycles, for example. They can also be counter-intuitive.

Drug treatment makes virus dynamics linear. Bonhoeffer and Nowak wrote down the basic equation for virus dynamics under drug treatment and solved it. The result was surprising at first sight. The observed slope of free virus decay can either reflect the death rate of free virus particles or the death rate of those cells that produce the virus, depending which of the two rates is slower. If infected cells die (are removed) slower than free viruses, then the slope is determined by the turnover rate of infected cells. If free virus particles die slower then the slope is determined by the turnover rate of free virus. A slope of 0.5 per day implies an average life-time (of free virus or virus-producing cells, whichever has the longer life-time) of 2 days.

It seems natural to assume that plasma virions have a shorter life-time than virus-producing cells. Therefore the observed decay kinetics of free plasma virus indicates the turnover rate of virus-infected cells. This was perhaps the main insight that the mathematical models offered for this study.

Those infected cells that produce (most of) plasma virus in HIV-1 infected patients have an average life-time of about 2–3 days. The half-life is the time it takes until 50% of such cells have died. The half-life equals the average life-time multiplied by the natural logarithm of 2 (which is about 0.693). If the average life-time is 2 days, then the half-life is about 1.4 days.

These findings imply a rapid turnover of infected cells in HIV infection. Most of the cells that produce plasma virus at any one time have only become infected within the last few days. HIV's long battle with the immune system is not a static affair. The idea that disease progression takes many years because not much is happening has been proven wrong.

This was an important finding, but it was not unexpected. For many years, people interested in HIV evolution had been witnessing enormous rates of turnover of different virus mutants on fast time-scales. Such a rapid genetic change is only compatible with a short generation time.

In our opinion the most important contribution of the new study was that it was possible to obtain a *quantitative* answer to a question concerning the dynamics of an infectious disease in its natural host: the half-life of virus-producing cells is about 2 days. The purely qualitative understanding of virus (or other) infections, in terms of molecular or cellular interactions, is now accompanied by a precise result concerning the dynamics. The rate of virus turnover in infected people can be measured. It is easy to see how the same method can be extended to other virus infections (and other pathogens)—as long as there is a drug to inhibit their replication.

But not all HIV-infected cells are short-lived. Already in the 1995 *Nature* paper, we noted that in fact most of the peripheral blood mononuclear cells (PBMC) that carry HIV provirus must be long-lived. Monitoring the rise of resistant virus mutants, we observed that it took only 2–4 weeks for the resistant mutant to reach 100% abundance in the plasma virus population, while it needed about 100 days for the mutant to reach 50% abundance in the PBMC provirus population. This implies that the majority of infected cells must be long-lived, with an average half-life of about 100 days.

How is it possible to reconcile the two different observations: data on plasma virus decay suggest that infected cells have a half-life of 2 days, whereas data on HIV provirus suggest they have a half-life of about 100 days? The answer is that the 2 day half-life refers to those cells that produce plasma virus, but most of the cells that carry HIV provirus do not produce plasma virus (or produce only very small amounts). We estimated that about 90% of the infected cells harbour HIV provirus but do not contribute significantly to plasma virus production. Many of them will have defective viral genomes that are incapable of completing the virus life-cycle.

In addition, there is a subset of infected PBMC that carry latent virus, which can be reactivated at a later time. We estimated the half-life of those cells between 10 and 20 days.

By 1995 a number of potent anti-HIV drugs were available, which could lead to a significant reduction in plasma virus load. The benefit of treatment, however, was only short-lived because the virus managed to come up with resistant mutants against all individual drugs. Virus load would decline initially, but return to pre-treatment levels after some weeks (or months). Following a long tradition of anti-microbial therapy, the obvious next step was combination treatment, using several drugs simultaneously, thereby hoping to outmanoeuvre the flexibility of the virus. A combination of two reverse transcriptase inhibitors, zidovudine and lamivudine, led to a 10-fold decline in plasma virus load which could be maintained for the duration of treatment.

Remarkable success came when three drugs were used at the same time. Various clinical trials demonstrated a strong and long-lasting effect of triple-drug therapy using two reverse transcriptase and one protease inhibitor. Plasma virus load could be observed to go down for several weeks, disappear below the detection limit (about 50 copies of viral RNA per ml) and remain below the detection limit in many patients for the duration of treatment. A new era of HIV treatment had began. For the first time new hope was dawning for millions of HIV-infected people. David Ho, a pioneer of anti-HIV combination therapy and a main proponent of the idea of hitting the virus 'hard and early' was voted the Time Magazine Man of the Year. It was 1996.

This chapter is organized as follows: in Section 4.1 we will explain the mathematical theory, describing the action of reverse transcriptase inhibitors, protease inhibitors, the rise of CD4 cells and the dynamics of long-lived infected cells; in 4.2 we discuss the clinical and experimental observations describing the short-term decay of plasma virus, the CD4 cell increase, the turnover of HIV provirus, the initial shoulder phase of plasma virus decline, the dynamics of triple-drug therapy and finally the question whether it is possible to eradicate HIV from infected patients.

4.1   Theory

4.1.1   *HIV: reverse transcriptase inhibitors*

In HIV infection, reverse transcriptase inhibitors prevent infection of new cells (Fig. 4.1). Suppose first, for simplicity, that the drug is 100% effective. Then we put $\beta = 0$ in eqn (3.1), and the subsequent dynamics of infected cells and free virus are given by

$$\dot{y} = -ay,$$
$$\dot{v} = ky - uv. \tag{4.1}$$

This leads to

$$y(t) = y^* e^{-at},$$
$$v(t) = \frac{v^*(u e^{-at} - a e^{-ut})}{u - a}. \tag{4.2}$$

Infected cells fall purely as an exponential function of time, whereas free virus falls exponentially after an initial 'shoulder phase'. If the half-life of free virus particles is significantly shorter than the half-life of virus-producing cells, $u \gg a$, then (as illustrated in Fig. 4.2) plasma virus abundance does not begin to fall noticeably until the end of a shoulder phase of duration $\Delta t \approx 1/u$ (more precisely, $\Delta t = -(1/a)\ln(1 - a/u)$).

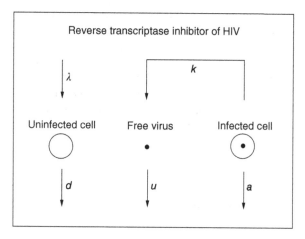

Fig. 4.1 A reverse transcriptase inhibitor of HIV prevents free virus particles from infecting cells: a virion which invades a cell cannot complete its reverse transcription step; the cell remains uninfected. The basic model of virus dynamics falls into two unconnected parts, one describing the dynamics of uninfected cells, the other describing the decay of infected cells and free virus.

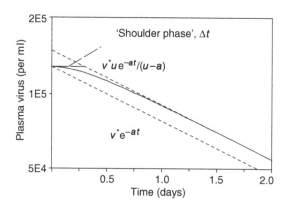

Fig. 4.2 Following initiation of therapy, plasma virus is expected to decline exponentially after an initial shoulder phase. Provided infected cells live longer than free virus, the length of the shoulder phase is approximately given by the average life-time of free virions, $1/u$, and the slope of the exponential decay by the death rate, a, of productively infected cells.

Thereafter virus decline moves into its asymptotic phase, falling as e^{-at}. Hence, the observed exponential decay of plasma virus reflects the half-life of virus-producing cells, while the half-life of free virus particles determines the length of the shoulder phase (Fig. 4.2).

Note that the equation for $v(t)$ is symmetric in a and u. Therefore, if a were larger than u, then the shoulder phase would be of duration $1/a$ and the subsequent exponential decline of plasma virus would be e^{-ut}. But we know that this case does not apply to HIV.

In the more general case when reverse transcriptase inhibition is not 100% effective, we may replace β in eqn (3.1) with $\bar{\beta} = s\beta$, with $s < 1$ (100% inhibition corresponds to $s = 0$). If the time-scale for changes in the uninfected cell abundance, $1/d$, is longer than other time-scales ($d \ll a, u$), then we may approximate $x(t)$ by x^*. We find the amount of free virus, $v(t)/v^*$, declines asymptotically as

$$[\Lambda_2/(\Lambda_2 - \Lambda_1)]\exp(-\Lambda_1 t),$$

where

$$2\Lambda_{1,2} = (u + a) \mp \sqrt{(u - a)^2 + 4sau}$$

(providing $d \ll a, u$). Thus the asymptotic slope of $\ln[v(t)]$ versus t gives a measure of Λ_1, while the duration of the 'shoulder phase', Δt, can be assessed from $\Delta t = (1/\Lambda_1)\ln[\Lambda_2/(\Lambda_2 - \Lambda_1)]$. In the limit $s \to 0$ (drug 100% effective), we have $\Lambda_1 = a$ and $\Delta t = (1/a)\ln[u/(u - a)]$, for $u > a$. More generally, for positive $s < 1$ and $u \gg a$, we see that the asymptotic decay rate is approximately

$$\Lambda_1 \approx a(1 - s)\left(1 - \frac{sa}{u} + O\frac{a^2}{u^2}\right)$$

and the shoulder phase duration is correspondingly

$$\Delta t \approx \left(\frac{1}{u}\right)\left(1 + \frac{a}{2u}(1 - 3s) + O\frac{a^2}{u^2}\right).$$

It follows that the decline in free virus abundance is still described by Fig. 4.2, except now the asymptotic rate of decay is $\exp[-at(1 - s)]$ for $u \gg a$; the duration of the shoulder phase remains $\Delta t \approx 1/u$. Thus the observed half-life of virus-producing cells, $T_{1/2} = (\ln 2)/[a(1 - s)]$, depends on the efficacy of the drug. Therefore, we expect that a more effective drug leads to a faster decay of plasma virus.

4.1.2 HIV: protease inhibitors

Protease inhibitors of HIV, prevent infected cells from producing infectious virus particles. Free virus particles, which have been produced before therapy starts, will for a short while continue to infect new cells, but infected cells will produce non-infectious virus particles, w (Fig. 4.3). The equations become

$$\dot{y} = \beta x v - ay,$$
$$\dot{v} = -uv, \tag{4.3}$$
$$\dot{w} = ky - uw.$$

The situation is more complex, because the dynamics of infected cells and free virus are not decoupled from the uninfected cell population. However, we can obtain analytic

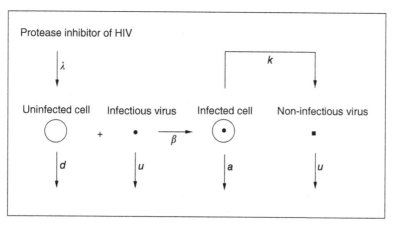

Fig. 4.3 A protease inhibitor of HIV prevents infected cells from producing infectious virions; instead they produce non-infectious virions. Once the infectious virions present at the start of therapy have disappeared (which happens rapidly), the system behaves in the same way as for a reverse transcriptase inhibitor treatment.

insights if we assume that the uninfected cell population remains roughly constant for the time-scale under consideration. In this case, we obtain

$$y(t) = \frac{y^*(ue^{-at} - ae^{-ut})}{u - a},$$

$$v(t) = v^*e^{-ut},$$

$$w(t) = v^*\left[(e^{-at} - e^{-ut})\frac{u}{u - a} - ate^{-ut}\right]\frac{u}{u - a}. \tag{4.4}$$

Infectious virions, $v(t)$, fall exponentially as e^{-ut}. Their decay slope provides a direct estimate for the turnover rate of free virus particles. Infected cells, $y(t)$, fall exponentially as e^{-at} after an initial shoulder phase of duration $1/u$ (provided $u \ll a$). The change in total abundance of free virus is given by

$$v(t) + w(t) = v^*\left[e^{-ut} + \left\{(e^{-at} - e^{-ut})\frac{u}{u - a} - ate^{-ut}\right\}\frac{u}{u - a}\right]. \tag{4.5}$$

Again, for $u \ll a$ this function describes a decay curve of plasma virus with an initial shoulder (of duration $\Delta t = -(2/a)\ln(1 - a/u) \approx 2/u$) followed by an exponential decay as e^{-at}. The situation is very similar to reverse transcriptase inhibitor treatment. The main difference is that the virus decay function is no longer symmetric in u and a, and therefore, a formal distinction between these two rate constants can be possible.

If, on the other hand, $a > u$, the asymptotic behaviour is no longer simply exponential, but rather $v(t)/v^* \to [a/(a - u)]ut\,e^{-ut}$.

4.1.3 Rise of uninfected cells

If a drug prevents the infection of new cells ($\beta = 0$) then from the basic model of virus dynamics, we obtain for the uninfected cell population:

$$\dot{x} = \lambda - dx. \tag{4.6}$$

Subject to the initial conditions x^* at $t = 0$, this gives rise to the following solution:

$$x(t) = \frac{\lambda}{d} - \left(\frac{\lambda}{d} - x^*\right)e^{-dt}. \tag{4.7}$$

According to this model, uninfected cells start to rise as a linear function of time. The initial slope is simply $\lambda - dx^*$. Thus for the same value of λ and d, patients with low x^* should have the higher initial increase. The rise of uninfected cells in the absence of viral resurgence would eventually level out converging to the uninfected equilibrium, $x = \lambda/d$.

4.1.4 Long-lived infected cells

In HIV infection, there are different types of infected cells (Fig. 4.4). In the previous sections we have only considered short-lived, virus-producing cells. In fact such cells produce most (more than 99%) of plasma virus. But in addition, there are

  (i) latently infected cells, which do not produce new virions, but harbour replication-competent virus: these cells can be reactivated to become virus-producing cells;

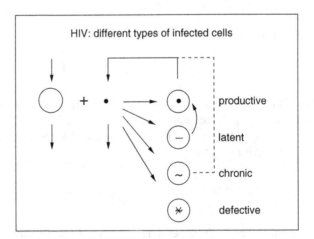

Fig. 4.4   There are different types of HIV infected cells: productively infected cells are engaged in producing new virions; latently infected cell can be reactivated to become productively infected cells; chronically infected cells may produce virions at a slow rate and live for a long time; cells harbouring defective provirus are unable to contribute to virion production.

(ii) long-lived chronic producers, which may produce small amounts of virus particles over a longer period; and

(iii) cells which harbour defective provirus.

We will therefore, expand the basic model of virus dynamics to include long-lived cells. Let us consider a model with virus-producing cells, y_1, latently infected cells, y_2, and cells harbouring defective virus, y_3. We obtain the following system of differential equations:

$$
\begin{aligned}
\dot{x} &= \lambda - dx - \beta xv, \\
\dot{y}_1 &= q_1 \beta xv - a_1 y_1 + \alpha y_2, \\
\dot{y}_2 &= q_2 \beta xv - a_2 y_2 - \alpha y_2, \\
\dot{y}_3 &= q_3 \beta xv - a_3 y_3, \\
\dot{v} &= ky_1 - uv.
\end{aligned}
\tag{4.8}
$$

Here, q_1, q_2, and q_3 denote, respectively, the probabilities that upon infection a cell enters the virus-producing state, the latent state, or obtains defective provirus. Clearly $q_1 + q_2 + q_3 = 1$. The death rates of these infected cell types are a_1, a_2, and a_3. We expect a_1 to be much larger than a_2 and a_3. Latently infected cells become reactivated to turn into virus-producing cells at rate α. We expect a_1 to be larger than $a_2 + \alpha$, which in turn should be larger than a_3.

The equilibrium of (4.8) is given by

$$
\begin{aligned}
x^* &= x_0/R_0, \\
y_1^* &= (R_0 - 1)\frac{du}{\beta k} = v^* \frac{u}{k}, \\
y_2^* &= y_1^* \left(\frac{a_1}{q_1}\right) \bigg/ \left(\frac{\alpha + a_2}{q_2} + \frac{\alpha}{q_1}\right), \\
y_3^* &= y_2^* \left(\frac{\alpha + a_2}{q_2}\right) \bigg/ \left(\frac{a_3}{q_3}\right), \\
v^* &= (R_0 - 1)\frac{d}{\beta}.
\end{aligned}
\tag{4.9}
$$

Here $x_0 = \lambda/d$ is the equilibrium abundance of uninfected cells prior to infection, and the basic reproductive ratio is given by

$$
R_0 = \frac{\beta \lambda k}{a_1 du}\left(q_1 + q_2 \frac{\alpha}{\alpha + a_2}\right).
\tag{4.10}
$$

Let us consider an anti-viral therapy, which prevents infection of new cells and, therefore, reduces β to zero. The decline of infected cells and free virus is given by the simple set of linear equations

$$\dot{y}_1 = -a_1 y_1 + \alpha y_2,$$
$$\dot{y}_2 = -a_2 y_2 - \alpha y_2,$$
$$\dot{y}_3 = -a_3 y_3,$$
$$\dot{v} = k y_1 - u v.$$

(4.11)

This leads to

$$y_1 = y_1^* e^{-a_1 t} + \frac{\alpha y_2^*}{a_1 - (a_2 + \alpha)} (e^{-(a_2+\alpha)t} - e^{-a_1 t}),$$
$$\dot{y}_2 = y_2^* e^{-(a_2+\alpha)t},$$
$$\dot{y}_3 = y_3^* e^{-a_3 t},$$
$$\dot{v} = v^* e^{-ut} + \frac{k}{u - a_1} \left(y_1^* - \frac{\alpha y_2^*}{a_1 - (a_2 + \alpha)} \right) (e^{-a_1 t} - e^{-ut})$$
$$+ \frac{k \alpha y_2^*}{(u - [a_2 + \alpha])(a_1 - [a_2 + \alpha])} (e^{-(a_2+\alpha)t} - e^{-ut}).$$

(4.12)

The free virus, v, falls in three different phases under the assumption that $u \gg a_1 \gg a_2 + \alpha$. Initially there is an interval of duration around $1/u$, during which $v(t)$ does not change; on this fast time-scale, $y_1(t)$ is unchanging, and so v remains at its equilibrium value. On the slow time-scale around $1/a_1$, there is a 'shoulder' phase, during which $v(t)$ declines to around

$$v^* \Big/ \left[1 + \frac{q_1}{q_2} \left(\frac{\alpha + a_2}{a_2} \right) \right].$$

Finally there is a phase of exponential decline, at the rate $a_2 + \alpha$, corresponding to a characteristic time-scale of $1/(a_2 + \alpha)$ (Fig. 4.5).

4.2   Experiment

4.2.1   Short-term decay

For the 1995 Nature paper, George Shaw, working with clinician Michael Saag, treated 22 HIV-1 infected patients with one of three drugs: two protease inhibitors (ritonavir and L-735,524) and the reverse transcriptase inhibitor, nevirapine (Fig. 4.6). David Ho and his clinician, Martin Markowitz, treated 20 HIV-1 infected patients with the protease inhibitor ritonavir. In all patients virus load was approximately constant in the week prior to treatment. Following the initiation of anti-viral therapy, there was a rapid decline of plasma virus. In most patients virus load declined 10- to 100-fold during the first 2 weeks of therapy. Fitting simple exponential decay curves provided average rates of plasma virus decline of about 0.5 per day. Assuming that the decay rate of free virus is

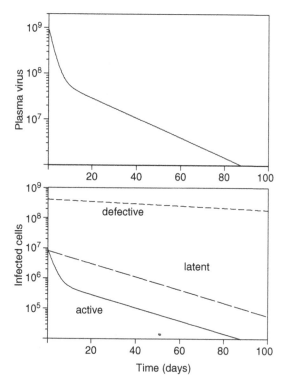

Fig. 4.5 Long-term dynamics of plasma virus and infected cell decline under effective therapy as described by model (4.8) and eqn (4.12). Plasma virus is produced by active cells, latently infected cell can turn into active cells, defectively infected cells do not contribute to viral reproduction. Parameter values are $\lambda = 10^7$, $d = 0.1$, $a_1 = 0.5$, $a_2 = 0.01$, $a_3 = 0.008$, $\alpha = 0.4$, $\beta = 5 \times 10^{-10}$, $q_1 = 0.55$, $q_2 = 0.05$, $q_3 = 0.4$, $k = 500$, $u = 5$.

much faster than the decay rate of virus-producing cells, the observed decay in plasma virus characterizes the kinetics of infected cells. Therefore, the experimental results suggest that virus-producing cells in HIV-1 infected patients have an average life-time of about 2.7 days, which is equivalent to a half-life of 1.9 days. In other words, about 30% of those infected cells are replenished every day.

After about 2 weeks of therapy, virus decline came to a stop. The residual viremia can be the consequence of virus production from long-lived infected cells, gradual activation of latently infected cells, inadequate drug concentration in certain tissues or viral resistance. In some patients, virus load started to rise again as a consequence of resistant mutants.

The treated patients differed greatly in the amount of plasma virus before therapy ranging from 10^4 to about 10^7 virus particles per ml. There was no correlation between the decay rate of virus-producing cells and baseline virus load.

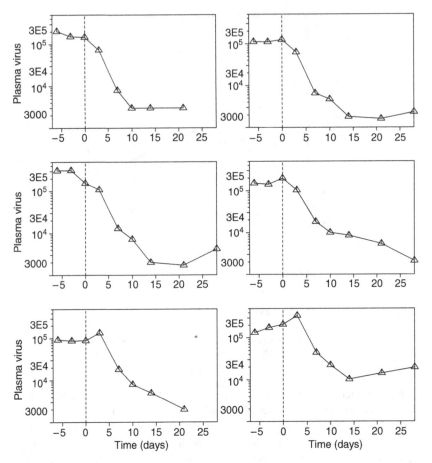

Fig. 4.6 Plasma virus load declines rapidly during treatment with the protease inhibitor ritonavir. Six representative patients are shown. Treatment starts at day 0. Plasma virus (copies HIV RNA per ml) falls about 100-fold in two weeks. The data are from Wei *et al.* (1995).

In a different study, also published in Nature in 1995, Clive Loveday and colleagues analysed 11 patients treated with the reverse transcriptase inhibitor zidovudine and obtained similar results. Virus load declined rapidly and reached a nadir about 6–14 days after starting therapy. The average rate of virus decline was 0.46 per day corresponding to a half-life of 1.5 days. This implies that similar insights into virus dynamics can also be obtained with zidovudine therapy, which had been in use for several years.

4.2.2 CD4 cell increase

Before therapy, patients had baseline CD4 cell counts ranging from 10 to 500 per μl. Concomitant with the decline in virus load during therapy was an increase in CD4 cell numbers. In some patients the increase was dramatic, in others only modest. The average rate of increase was about 8 cells per μl per day. The average total increase was about 200 cells per μl. Patients with a lower CD4 cell count had a higher rate of increase. It could not be distinguished whether the rise was exponential or linear. An exponential increase would be caused by proliferation of CD4 cells, a linear increase could be the consequence of the production of CD4 cells from a precursor source such as the thymus.

The baseline CD4 cell values and the rates of increase can be used to estimate the total numbers of CD4 cells eliminated by HIV every day. In Shaw's patients, the maximum increase of CD4 cell numbers ranged from about 40 to 800 cells per μl, with an average of about 200 cells per μl. Assuming a constant production of CD4 cells from a pool of precursor cells the average increase was 8 cells per μl per day. Given that the peripheral blood contains only 2% of the total body lymphocytes and that the average blood volume is roughly 5 litres, this implies that about 2×10^9 CD4 cells are being produced per day. In other words, for maintaining a steady-state prior to treatment the virus must have eliminated 2×10^9 CD4 each day.

There are potential problems with the above estimate. First, we assumed that the rate of increase of CD4 cell numbers in the blood reflects the rate of increase in the lymph system. This need not be true. Second, a declining virus load in the lymph system can lead to a release of CD4 cells from the lymph system into the blood. Thus the observed increase of CD4 cells in the blood can in part be the consequence of a redistribution of cells from the lymph system into the blood. Further experiments are necessary to evaluate these possibilities.

Another interesting observation was that the rate of virus decline was independent of the CD4 cell count of the patients prior to treatment. One could have expected that patients with a higher CD4 cell count may have a more effective anti-viral CTL response and should thus have a shorter half-life of infected cells, which constitute targets for CTL-mediated lysis. We will come back to this point in Chapter 6.

4.2.3 Long-lived infected cells

In three patients treated with the reverse transcriptase inhibitor nevirapine, George Shaw monitored the rise of resistant virus mutants (Fig. 4.7). In all patients, the resistant virus, which is characterized by a single point mutation in the reverse transcriptase gene, reached 100% frequency in the plasma virus RNA population within 4 weeks of initiating therapy. In the provirus DNA population of peripheral blood mononuclear cells (PBMC), however, it took 20 weeks for the mutant to reach 50–80% prevalence. This implies that the HIV provirus population in PBMC turns over at a very slow rate compared to the plasma virus population. In fact, the above data suggest that the half-life of HIV infected PBMC is about 100 days. Thus, the half-life of infected PBMC is very long and of the same order of magnitude as the half-life of uninfected PBMC, suggesting that most of the infected cells harbour defective viral genomes and that the plasma virus population is generated by only a small portion of infected cells that have a half-life of about 2 days.

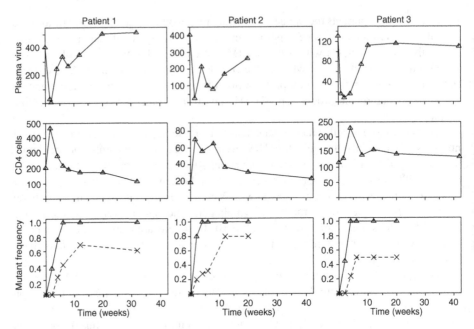

Fig. 4.7 Three patients receiving nevirapine (NVP) treatment. The figures shows changes over time in plasma virus load (copies of HIV RNA per µl), CD4 cell numbers (per µl) and the frequency of NVP resistant mutant in plasma virus RNA population (continuous line) and PBMC DNA population (dashed line). The data are from Wei *et al.* (1995).

In a subsequent paper (Nowak *et al.* 1997*a,b*), we showed that about 90% of the infected PBMC have essentially defective genomes. The remaining 10% of cells consist of rapidly turning over cells with a half-life of 2 days and latently infected cells with a half-life of 10–20 days. Latently infected cells can be stimulated *in vitro* to give rise to new virions.

In vivo, about 99% of plasma virus present in an untreated patients comes from the short-lived cells that have a half-life of 2 days. This is a consequence of the observation that plasma virus load declines about 100-fold with a constant half-life of 2 days.

4.2.4   *Virion turnover*

In a subsequent study, David Ho and his colleagues treated five infected patients with the protease inhibitor ritonavir and sampled plasma virus load at very frequent intervals: every 2 h until the sixth hour, every 6 h until day 2, and every day until day 7 (Fig. 4.8). In all patients, there was an initial shoulder followed by an exponential decline. The slope of the exponential decay was on average 0.5 per day confirming the earlier estimates which were based on a much less frequent sampling. In addition, the initial shoulder could be used to obtain a minimum estimate for the clearance rate of free virus particles

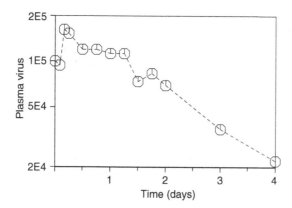

Fig. 4.8 Very frequent sampling during the first 2 days of treatment outlines the shoulder phase prior to the exponential decay of plasma virus. The patient received the protease inhibitor ritonavir. Plasma virus is shown in copies per ml. The data are from Perelson *et al.* (1996).

(using eqn (4.5)). The average clearance rate was about 3.1 per day leading to a half-life of about 5 h (Perelson *et al.* 1996).

The minimum estimate of the virus clearance rate can be used to calculate a minimum estimate of virus production prior to treatment. For example, in one patient plasma virus load before therapy was about 3×10^5 virions per ml. The clearance rate was estimated to be 3.81 per day. This means that about 2.9×10^5 virions per ml are cleared per day. The same amount must have been produced prior to treatment for virus load to be at steady-state. For 3 litres of plasma this would come to about 10^9 virions per day. On average, Perelson and Ho estimated that about 10^{10} virions are produced in an HIV infected individual per day.

Both the figure for the clearance rate of virions and their total daily production are minimum estimates, because the length of the shoulder phase is also influenced by the intracellular time of the virus life cycle (Herz *et al.* 1996). Furthermore, in essentially all patients virus load increased in the first hours after initiating therapy. The reasons for this are unknown. This initial increase in virus load may artificially extend the length of the shoulder phase and therefore confound the estimates for the decay rate of free virus.

4.2.5 *Triple-drug therapy*

During successful combination therapy plasma virus load continues to decline for several weeks. Alan Perelson and David Ho analysed data from eight patients receiving triple-drug therapy. Initially virus load declined with a half-life of about 1.1 days, which was slightly faster than earlier observations, perhaps because triple-drug therapy was more potent than single-drug therapy. Subsequently, the decay slows down and occurred with an average half-life of about 2 weeks (ranging from 1 to 4 weeks). In addition, they measured the decay rate of infected PBMC that contained infectious virus and obtained an average half-life of about 1 week (with a range from 0.5 to 2 weeks). The decline of proviral DNA occurred with an average half-life of 20 weeks (ranging from 3 to 62 weeks).

Perelson and Ho concluded that the latently infected PBMC did not contribute to the second phase of plasma virus decline, because the half-lives of the two decays were different. Instead, they conjectured that the second phase of plasma virus decay must be the consequence of another long-lived infected cell population; for example infected macrophages.

They also estimated the minimum time it would take for triple-drug therapy to eliminate the longest-lived infected cell population that contained infectious virus. If the longest-lived population decays strictly exponentially with rate a_{min}, and if the size of this population is N, then the time for eradication is given by

$$t = (1/a_{min}) \ln N. \tag{4.13}$$

Using the half-life of the second phase of plasma virus decay with a range of 1–4 weeks and assuming that there are (at most) 10^{12} long-lived infected cells, the minimum time for virus eradication would be 0.8–3.1 years.

4.2.6   Eradication

The overwhelming success of triple-drug therapy has now led to the question whether it is realistic to eradicate HIV from infected patients and for how long treatment has to continue to achieve this aim.

The above estimate of 0.8–3.1 years is based on an extrapolation of an exponential decay curve that has been observed for 6 weeks. It is unlikely, however, that the long-lived infected cell population that is responsible for this exponential decay will continue to decay with the same rate for several years. If this cell population is heterogeneous, due to variation in viral or cellular properties, then it is more likely that the decay rate will slow down over time, because the shorter-lived subpopulations will disappear earlier, while longer-lived cells will die later. Hence, the expectation is that sampling at a time later than 6 weeks after initiating therapy will lead to a longer estimate of the half-life of these infected cells.

Such an experiment was performed by Tae-Wook Chun and Anthony Fauci at the National Institute of Health in Bethesda. Using a more sensitive method to quantitate infected cells, they could show that in 13 of 13 patients receiving triple-drug therapy for an average duration of 10 months there were cells containing infectious HIV provirus (Chun et al. 1997a).

The data from this study can be used to obtain a rough estimate of the half-life of latently infected PBMC for a measurement obtained 10 months after initiating therapy. In nine untreated patients, the average infected cell load is about 360 infectious units per million cells (IUPM). In nine treated patients with undetectable plasma virus, the average infected cell load is about 3.2 IUPM. The patients on triple-drug therapy have been receiving treatment for 7–12.5 months with an average of 10 months. Assuming that the infected cell load in the untreated patients is representative of the infected cell load in the treated patients prior to treatment, we derive a rough estimate for the half-life of infected resting CD4+ cells of about 6 weeks. This estimate is longer than the previous estimate of 1–4 weeks and suggests a minimum time for eradication of 4.6 years.

But again the same argument as above applies: it is likely that we have not yet seen the slowest subpopulation, the longest-lived cell. The expectation is that data sampled from patients after 2 years of triple-drug therapy will again lead to an increased estimate for the time to eradication.

Suppose it is possible for a replication competent HIV provirus to remain in a completely latent state with virus protein expression so low (or absent) that the cell is not a target for CTL-mediated lysis. In this case the half-life of such a cell should be equivalent to the half-life of a cell harbouring defective provirus. Thence the relevant time-scale for virus eradication is defined by the decay rate of HIV provirus. Taking a 20 week half-life, the time for eradication increases to 15 years. It is also possible that in certain patients there is a low but continuous level of infection of new cells, which could lead to a stable, non-declining level of long-lived infected cells.

Therefore, we believe that complete eradication is not a practical option with current anti-viral treatment. A tremendous achievement would be combining triple-drug therapy with agents that activate CD4 cells and, therefore, activate HIV provirus, thereby reducing the life-time of latently infected cells. This could greatly shorten the time for eradication of HIV in infected people. Activating CD4 cells is, however, not unproblematic, especially since a large fraction of CD4 cells have to be activated in order to get those few that carry infectious HIV provirus.

4.3   Further reading

The original virus dynamics papers are Wei *et al.* (1995) and Ho *et al.* (1995). Coffin (1995) discusses the implications of the observed virus dynamics on genetic variation and HIV pathogenesis. Nowak *et al.* (1995) estimate parameters of virus dynamics from patients receiving AZT therapy. Perelson *et al.* (1996) analyse the short-term kinetics of virus decline taking hourly samples after the initiation of therapy. Herz *et al.* (1996) provide a mathematical description of the initial shoulder phase and the effect of the intracellular phase of the virus life cycle. Nowak *et al.* (1997*a,b*) estimate half-lives of latently infected cells. Perelson *et al.* (1997*a,b*) analyse data from triple-drug therapy and estimate the minimum time to eradication of HIV-1 infection. Bonhoeffer *et al.* (1997*a*) is a review of virus dynamics. Chun *et al.* (1997*a,b*, 1999), Finzi *et al.* (1997, 1999), Wong *et al.* (1997), and Chun and Fauci (1999) study the decay of latently infected cells. See also Bukrinsky *et al.* (1991). For quantification of HIV in tissue we refer to Haase *et al.* (1996) and Hockett *et al.* (1999). Ogg *et al.* (1999) study CTI decline during anti-HIV therapy. Kilby *et al.* (1998) describe a peptide inhibitor of virus entry.

5

DYNAMICS OF HEPATITIS B VIRUS

Hepatitis is a general term denoting infection or inflammation of the liver. The most notable sign of liver disease, jaundice, is already described in ancient medical sources such as the Babylonian Talmud or the writings of Hippocrates. There are several different viruses that can cause hepatitis. In this chapter, we focus on the hepatitis B virus (HBV).

Currently about 300 million people worldwide are chronically infected with the HBV. Upon encountering the virus, adults have a roughly 90% chance of undergoing an acute infection with clinical symptoms and subsequent clearance of the virus. Children, however, have usually mild or no clinical symptoms immediately after infection, but do not clear the virus; they often become chronic carriers. About 25% of chronic carriers will die from liver cancer induced by the virus. Thus HBV is the most common viral source of human cancer worldwide and the second most important human carcinogen in general, beaten only by tobacco. The virus is transmitted by contact with saliva and blood products, including venereal and mother to infant transmissions. Early vaccination programs with unsterilized needles also led to the spread of HBV.

HBV was discovered in the 1960s by Baruch Blumberg, who was then studying genetic differences in human populations. He isolated a protein from Australian aborigines that cross-reacted with antibodies from the sera of American hemophilia patients. This 'Australian antigen' turned out to be the envelope protein of HBV.

HBV belongs to the family of hepadnaviruses. It has a circular DNA genome which is about 3400 nucleotides long (1/3 the size of the HIV genome). Curiously, the HBV genome is partly double stranded and partly single stranded; the second strand is only 1700–2800 nucleotides long. The HBV particle is spherical with a diameter of about 45 nm. Surrounding the viral genome and a viral encoded DNA polymerase, is a protein coat which consists of the hepatitis B core antigen (HBcAg). This core is surrounded by an envelope membrane, which derives from the host cell and contains in addition the hepatitis S antigen (HBsAg). The virus genome encodes four proteins: the DNA polymerase, the core antigen, the S antigen and another protein, called X protein, whose function seems to be the upregulation of viral mRNA production.

The HBsAg enables the virus to attach to a liver cell. After the virus has entered the cell, the viral polymerase completes the partially single stranded DNA into a complete double stranded DNA, called cccDNA (for covalently closed circular DNA). In the cell nucleus, this cccDNA is then transcribed by host enzymes into viral RNA, some of which act as mRNA for the production of viral proteins, while the full-length RNA acts as template for the production of new HBV DNA: the viral DNA polymerase also acts as a reverse transcriptase! This unusual replication cycle is unique to hepadnaviruses and makes them similar to retroviruses. In fact, hepadnaviruses can be seen as reciprocal retroviruses. They use a reverse transcription step, but carry the DNA genome in their virion.

Virus replication in the human host seems to be extremely efficient. In chronic carriers, there can be more than 10^9 virions per ml blood, which is more than 1000-fold the virus load usually seen in asymptomatic HIV patients. In addition the plasma of infected patients contains large amounts of incomplete virus particles which are only composed of viral envelope together with HBsAg.

A major breakthrough in fighting the HBV epidemics was the development of an effective vaccine in the late 1970s, using the HBsAg to stimulate effective antibody responses. While the vaccine can greatly reduce the risk of infection, it does not help those patients who are already infected.

Until very recently there was no effective treatment for chronic carriers of HBV. In the 1990s, however, it turned out that the reverse transcriptase inhibitor lamivudine, which was designed as anti-HIV drug, was also effective against HBV.

In this chapter, we analyse clinical data from HBV infected patients undergoing lamivudine treatment and perform an analysis similar to that for HIV in the previous chapter. Treatment of HBV infections with the reverse transcriptase inhibitor lamivudine leads to a rapid decline in plasma viraemia and provides estimates for the crucial kinetic constants of HBV replication. In persistently infected patients, HBV particles are cleared from the plasma with a half-life of about 1.0 days, which implies a 50% daily turnover of the free virus population in the absence of treatment. Total viral release into the periphery is around 10^{11} virus particles per day. Virus-producing cells have different half-lives in different patients, ranging from 10 to 100 days. Thus, although the rapid turnover of plasma virus (with enormous numbers of virions being produced every day for many years of persistent infection) is similar to HIV, the length and variability of the life-time of productively infected cells is in sharp contrast.

5.1 Theory

In the replication cycle of HBV, the reverse transcriptase is primarily responsible for the synthesis of new HBV DNA from the pre-genomic mRNA template. Therefore, a reverse transcriptase inhibitor can prevent the production of new virus particles from already infected cells (Fig. 5.1). We put $k = 0$ in eqn (3.1) and obtain for the dynamics of infected cells and free virus

$$\dot{y} = \beta x v - a y,$$
$$\dot{v} = -u v. \tag{5.1}$$

Again assuming that x is constant on the relevant time-scale, we obtain

$$y(t) = y^* \frac{(u e^{-at} - a e^{-ut})}{u - a},$$
$$v(t) = v^* e^{-ut}. \tag{5.2}$$

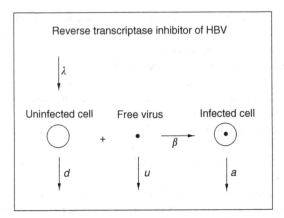

Fig. 5.1   A reverse transcriptase inhibitor of HBV prevents infected cells from producing free virions. Following infection of a cell, the viral DNA genome is transcribed
into RNA by a host cell enzyme. The RNA is then reversely transcribed into DNA
by the viral polymerase. This step is inhibited.

Hence, free virus falls purely exponentially as e^{-ut}, whereas infected cells fall exponentially as e^{-at} after an initial shoulder phase of duration $1/u$. From the decline of
free virus we can estimate the decay rate u. Before treatment, steady-state of free virus
implies $ky_0 = uv_0$. We know v_0, hence we can estimate ky_0, which is the total virus
production per day.

There is also some evidence that lamivudine blocks the completion of the double-
stranded circular DNA before migration to the cell nucleus. Hence, lamivudine could in
addition also prevent the infection of new cells: $\beta = 0$. This would lead to

$$\dot{y} = -ay,$$
$$\dot{v} = -uv. \tag{5.3}$$

In this case, both free virus and infected cells fall exponentially with time with rate
constants u and a, respectively:

$$y(t) = y^* e^{-at},$$
$$v(t) = v^* e^{-ut}. \tag{5.4}$$

Of course, the difference between eqns (5.2) and (5.4) is negligible if $u \gg a$. Thus,
if lamivudine does not effectively prevent infection of new cells, then (5.4) is still a very
good approximation of (5.2), provided u is substantially larger than a, which will turn
out to be true.

5.2   Experiment

In a clinical study, chronic HBV carriers were treated for 28 days with various doses
of lamivudine. Plasma virus load was quantified at frequent time points before, during

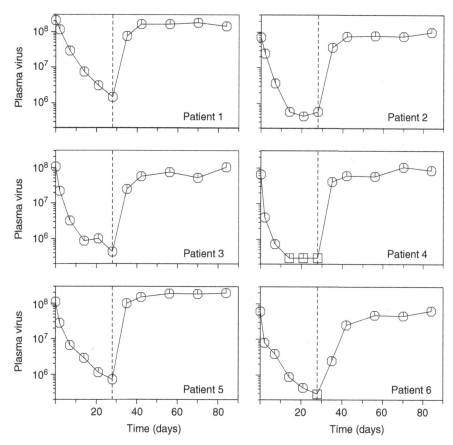

Fig. 5.2 Rapid decline in plasma virus in six patients receiving lamivudine. After 28 days, the treatment was stopped and virus load returned rapidly to pre-treatment levels. Plasma virus is shown in copies of viral DNA per ml. Squares indicate that virus load was below detection limit.

and after treatment. Figure 5.2 shows plasma virus changes in six patients. After the onset of therapy, viral levels decline rapidly, but as soon as the drug is withdrawn virus returns.

From this study we can estimate the initial rate of virus decline during the first two days of treatment. We obtain an average of $u = 0.67$ per day (standard deviation $\sigma = 0.32$, sample size $n = 23$), which corresponds to a half-life time ($T_{1/2}$) of 1.0 days. Hence, in the absence of treatment about 50% of the plasma virus is replenished every day. The total serum virus load (for 3 litres of serum) before treatment varies in different patients ranging from 10^{10} to 10^{12} particles with an average of 2.2×10^{11} ($\sigma = 2.6 \times 10^{11}$, $n = 45$). Consequently the total amount of virus production (release into the serum) follows a wide distribution with an average of 1.3×10^{11} particles per day ($\sigma = 8.2 \times 10^{10}$, $n = 23$). These differences are likely to reflect different population sizes of virus-infected cells in individual patients.

Virus decline during treatment is not strictly exponential (see Fig. 5.2). This can be explained by the assumption that the efficacy of the drug is not 100%. Assuming a certain efficacy ρ, viral decay occurs according to

$$v(t) = v_0(1 - \rho + \rho e^{-ut}). \tag{5.5}$$

Fitting this function provides an estimate for the efficacy of the drug at various doses. We find that for daily doses of 20, 100, 300 and 600 mg viral replication is inhibited by 87%, 97%, 96%, and 99%, respectively.

When therapy is withdrawn, virus resurges according to $dv/dt = ky - uv$. Hence, the initial virus growth rate can be approximated by

$$v(t) = v_1 e^{-ut} + \left(\frac{ky_1}{u}\right)(1 - e^{-ut}), \tag{5.6}$$

where v_1 and y_1 indicate the levels of free virus and infected cells at the end of therapy. We know v_1 by direct measurement, we have determined the decay rate, u, and hence we can estimate ky_1, which is the rate of virus production from infected cells at the end of therapy. Comparing ky_0 and ky_1 gives an estimate for the decay rate of infected (virus producing) cells, a, and consequently their half-life. This method requires the *initial* growth rate of virus after therapy has stopped, since y is treated as a constant. The approximation is accurate if virus load is determined early after the end of treatment. In our study, treatment was withdrawn after 28 days, and virus load was determined at days 28 and 35. For the different patients we obtain a broad distribution with an average decline of $a = 0.043$ per day ($\sigma = 0.035$, $n = 20$) corresponding to a $T_{1/2}$ of 16 days (ranging from 10 to 100 days).

The broad distribution of turnover rates of infected cells may be a consequence of heterogeneity of the immune response against infected cells in different patients. Damaged liver cells release ALT, and hence plasma ALT levels should provide some crude estimate for the amount of cell death in the liver. If most cell damage is caused by immune responses directed to infected cells, then ALT levels provide some estimate for the strength of the immunological response against HBV. We find a positive correlation between the decay rate of infected cells and the pre-treatment ALT level among different patients. This supports our claim that the variability of cell decay rates reflects different strengths of anti-cellular immune responses and is not simply caused by fluctuations in measurement or inaccurate approximation.

In a subsequent study, treatment continued for 24 weeks. Again we observe rapid decline in plasma virus load, which falls below detection limit in almost all patients within 2–4 weeks, and again in most patients virus resurges rapidly as soon as the drug is withdrawn. Figure 5.3 shows changes in plasma virus in six patients treated for 24 weeks. In addition, the plasma concentrations of HBeAg and ALT decline slowly during long-term lamivudine treatment.

HBeAg is produced by infected cells; its production is not directly inhibited by lamivudine, and changes in the serum concentration therefore reflect changes in infected liver cell mass. Thus, we can obtain an independent estimate of the turnover of infected

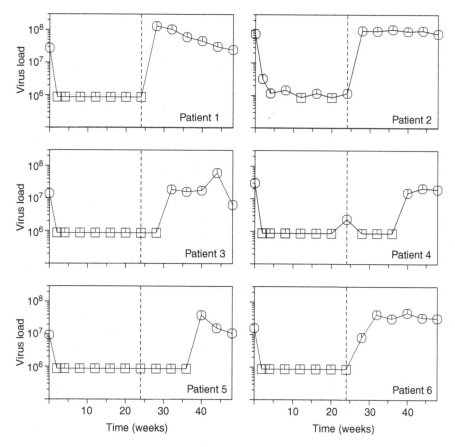

Fig. 5.3 HBV plasma virus load in six patients receiving lamivudine for 24 weeks. In most patients virus load declines rapidly below detection limit (square symbols). In all patients virus load returns rapidly to pre-treatment levels when treatment is stopped.

cells from the initial decay of HBeAg. During the first 8 weeks of treatment we observe an average decline of $a = 0.053$ per day ($\sigma = 0.039$, $n = 29$), which corresponds to a $T_{1/2}$ of 13 days. This is in excellent agreement with our previous estimate. Again we find a strong correlation between decay of infected cells and ALT levels among different patients.

After 24 weeks of treatment, patients were followed for another 24 weeks. Interestingly, there is a strong capacity in individual patients to return to the pre-treatment steady-state level after these 48 weeks. Plasma virus load, ALT levels, and HBeAg levels for each patient after 48 weeks are similar to their pre-treatment levels. Whatever factors (e.g. efficacy of anti-viral and anti-cellular immune responses) determine the particular pre-treatment steady-state level of plasma virus, HBeAg and ALT in individual patients, it is interesting to note that these factors have apparently not changed over the time course of 48 weeks in most patients.

It is likely that HBV infects and replicates at different rates in a number of cell types. Therefore the estimated viral production rates and turnover rates of infected cells have to be interpreted as average values.

A half-life of 10 days for infected, virus-producing hepatocytes implies that about 7% of this cell population is lost per day. In a chronic HBV infection between 5% and 40% of all hepatocytes can be infected and produce virus. Therefore, between 0.3% and 3% of all hepatocytes might be killed and must be replenished every day to maintain a stable liver cell mass. Since the liver contains around 2×10^{11} hepatocytes, this comes to 10^9 cells per day. This enormous activity of cell death and regeneration is likely to be a major driving force for development of hepatocellular carcinoma.

Frank Chisari from the Scripps Research Institute, however, has demonstrated that the CTL response in HBV infection can also have a *curative effect*. Instead of killing an infected cell, the CTL may release chemicals which reduce the HBV cccDNA copy number in the nucleus of the infected cell. Hence CTL can *cure* infected cells in the sense that they free them of virus without killing them. Therefore, during drug treatment the cccDNA copy number per infected cell may decline, which could lead to less viral and HBeAg production independent of death of infected cells. Hence, the decline of viral production and HBeAg could overestimate the actual death rate of infected cells and rather reflect the decay of cccDNA. What we measure is in fact the decay of potential virus 'production units' during treatment, independently of whether this reflects cell death or decay of viral cccDNA. Note, however, that patients with high ALT levels clear virus-producing units faster.

Treatment of chronic HBV infections with lamivudine leads to a rapid and sustained decline of plasma virus levels, but clinical benefit with a reduced risk of cirrhosis and development of liver cancer will greatly depend on the decline of infected cells. In patients where infected cells decline with a half-life of 10 days, treatment for one year could potentially reduce the number of infected cells to around 10^{-11} of its initial value. In patients with an infected cell half-life of 100 days, one year of treatment could reduce the number of infected cells to about 8% of its initial value. Thus lamivudine could be used over a prolonged period as single-agent therapy or to reduce the number of infected cells prior to immunotherapy designed to eradicate infected cells. Immunotherapy without anti-viral treatment could be problematic because of the very large number of infected liver cells in the typical HBV carrier. A quantitative understanding of HBV dynamics will make it possible to devise optimal treatment strategies for individual patients.

5.3   Comparing HBV and HIV

Both viruses can cause persistent infections which last for many years. In both infections, there is a rapid turnover of plasma virus. The currently best estimate for the half-life of HIV virions is about 6 h compared to about 24 h for HBV. Since, however, the average virus load in HBV is much higher, many more HBV than HIV virions are being produced per day. For 3 litres of plasma on average 10^{11} HBV particles are being produced and cleared every day, compared to 10^9 HIV particles.

Another crucial difference arises when we compare the half-lives of productively infected cells. In all HIV infected patients studied so far the half-life of productively infected cells is roughly the same, ranging from about 1–3 days, whereas in HBV infected patients the half-life of those cells varies greatly ranging from 10 to more than 100 days.

We suggest the following explanation for this difference between HIV and HBV. Most virologists believe that HBV is largely non-cytopathic *in vivo*, which means that it does not kill infected cells. The life-time of HBV infected cells is thus largely determined by the strength of the anti-HBV CTL response. Patients with a strong CTL response will clear infected cells rapidly, patients with a weak CTL response will clear them slowly. HIV, on the other hand, should be cytopathic *in vivo*. After some time, infected cells are killed by the virus, independent of CTL-mediated lysis. Hence, differences in the anti-HIV CTL response may only have a small effect on the observed half-life of infected cells. If HIV was non-cytopathic *in vivo*, as is sometimes suggested, then it is hard to explain why all patients sampled so far, have roughly the same (and very short) half-life of productively infected cells. In this case, all patients would have to have equally efficient CTL responses, which seems unlikely. We will return to this discussion in Chapter 7.

5.4 Further reading

Arias *et al.* (2000) is an excellent textbook on physiology and pathology of the liver. London and Blumberg (1982), Chisari and Ferrari (1995), and Payne *et al.* (1996) describe models for the pathogenesis of HBV infection. Chisari and Ferrari (1995) and Guidotti and Chisari (1996) are pertinent review articles. Also see Rehermann *et al.* (1996), Penna *et al.* (1996), Nakamoto *et al.* (1998), and Guidotti *et al.* (1994*a,b*, 1996*b*, 1999). The mechanism of lamivudine inhibition of HBV replication is described by Doong *et al.* (1991). Clinical trials of lamivudine are reported in Dienstag *et al.* (1994, 1995). HBV dynamics are analysed by Nowak *et al.* (1996*b*); hepatitis C virus dynamics are analysed by Neumann *et al.* (1998) and Cerny and Chisari (1999).

6

DYNAMICS OF IMMUNE RESPONSES

In this chapter, we analyse the basic mathematical models that describe the interaction between a replicating virus population and immune responses. We study a variety of such models exploring different assumptions about the detailed dynamics of the immune response. We characterize conditions that specify whether a virus will establish a persistent infection or will be cleared by the immune response. We also study the correlation between virus load (that is, the abundance of virus) and the magnitude of the anti-viral immune response. While functional immune responses always reduce virus load, the correlation between virus load and the abundance of immune cells can be positive or negative. A high virus load provides a strong antigenic stimulation and may induce a high abundance of immune cells which in turn reduce virus load.

The interactions between viruses and immune responses are reminiscent of predator–prey dynamics in classical ecology. Indeed the study of viral infections with a number of different cell types and different viral mutants can be seen as a kind of 'micro-ecology'. Viruses are prey, immune cells are predators. Immune cells 'feed' on viruses: they divide in response to contact with viral antigen and they kill virus.

An important concept in this chapter will be the 'immune responsiveness', which is defined as the rate at which an individual mounts an immune response to a given virus. In terms of a CTL response, 'CTL responsiveness' is the average rate at which specific CTL proliferate after encountering an infected cell. This rate will depend on factors such as the affinity of the T-cell receptor for the peptide–MHC complex, the level of T-cell help, the frequency of specific precursor cells and the concentration of cytokines. The CTL responsiveness against a specific virus is likely to vary among individuals and will depend, among other things, on the major histocompatibility complex (MHC) genes, which determine which epitopes of the virus are presented to the immune system. The CTL responsiveness of a patient can also vary over time, for example in HIV infection, as a result of declining T-cell help or antigenic variation. Emergence of antagonistic variants may also reduce CTL responsiveness.

In contrast to the inherent property of CTL responsiveness, 'CTL response' denotes the actual number of virus-specific CTL present at a given time. This is what is measured by *in vitro* assays. The CTL response depends on the amount of stimulation provided by the virus, and so on virus load. CTL response and virus load are linked to each other in a density-dependent fashion: a strong CTL response may reduce virus load, but the resulting low virus load will provide less stimulation, and in time the CTL response will decline.

In short, we argue that: (1) virus load is an important determinant of disease; (2) immune responses limit virus load; (3) individual variation in immune responsiveness accounts for much of the observed variation in the outcome of disease. The mathematical models in this chapter are designed to describe CTL responses. We concentrate on CTL because of their clear importance in the defence against virus infections. We point out, however, that similar principles will hold for other host defence factors, such as antibody responses.

6.1 A self-regulating CTL response

At first consider the simplest model for the interaction between a virus population and an immune response. The virus population replicates according to the basic model of virus dynamics (see Chapter 3). The immune response is triggered by encountering foreign antigen and then adopts a constant level which is independent of the concentration of virions or infected cells. Let us explore the effect of a CTL response, z, that eliminates infected cells:

$$
\begin{aligned}
\dot{x} &= \lambda - dx - \beta xv, \\
\dot{y} &= \beta xv - ay - pyz, \\
\dot{v} &= ky - uv, \\
\dot{z} &= c - bz.
\end{aligned}
\tag{6.1}
$$

There are four variables: uninfected cells, x, infected cells, y, free virus particles v and CTL, z. Infected cells are produced from uninfected cells and free virus at rate βxv and die at rate ay. Free virus is produced from infected cells at rate ky and declines at rate uv. Uninfected cells are produced at a constant rate, λ, and die at rate dx. CTL are produced at the rate c and die at rate bz. We assume that c is positive if $y > 0$, that is if an infection is present, otherwise $c = 0$. Infected cells are eliminated by the CTL response at the rate pyz.

As discussed earlier, the basic reproductive ratio of the virus in the absence of the CTL response is given by

$$
R_0 = \frac{\beta \lambda k}{adu}.
\tag{6.2}
$$

If $R_0 > 1$ then an infection can take place, and the immune response, z, will become activated. Whether or not the virus will establish a persistent infection depends on its basic reproductive ratio in the presence of the CTL response. This quantity is given by

$$
R_I = \frac{\beta \lambda k}{(a + a')du},
\tag{6.3}
$$

where $a' = cp/b$ is the rate at which infected cells are eliminated by the CTL response at its equilibrium level.

If $R_I < 1$ then the infection will be cleared. In this case, the virus may spread initially, but once the immune response is fully activated each infected cell will on average give rise to less than one newly infected cell. The virus population will decline and die out again.

If $R_I > 1$ then the infection will persist and the equilibrium abundances of cells, virions and CTL are given by

$$\hat{x} = \frac{(a + a')u}{\beta k},$$

$$\hat{y} = \frac{\lambda}{a + a'} - \frac{du}{\beta k},$$

$$\hat{v} = \frac{\lambda k}{(a + a')u} - \frac{d}{\beta}, \tag{6.4}$$

$$\hat{z} = c/b.$$

Without the CTL response, the system would converge to the equilibrium

$$x^* = \frac{au}{\beta k},$$

$$y^* = \frac{\lambda}{a} - \frac{du}{\beta k},$$

$$v^* = \frac{\lambda k}{au} - \frac{d}{\beta}, \tag{6.5}$$

$$z^* = 0.$$

Therefore, we see that the CTL response

- reduces the equilibrium abundance of infected cells;
- reduces the equilibrium abundance of free virions; and
- increases the equilibrium abundance of uninfected cells.

The total number of cells (infected plus uninfected), however, may be increased or reduced by a CTL response. We have

$$\hat{x} + \hat{y} = \frac{\lambda}{a + a'} + \frac{u(a + a' - d)}{\beta k}. \tag{6.6}$$

Let us consider the situation without CTL response, $a' = 0$. Figure 6.1 shows $x^* + y^*$ as a function of a. (i) If $a = d$ we have $x^* + y^* = \lambda/d$. This case corresponds to a non-cytopathic virus; infected cells have the same death rate as uninfected cells. (ii) If $a = \lambda\beta k/(ud)$ then again $x^* + y^* = \lambda/d$. This case corresponds to a virus with the highest possible level of cytopathicity, where $R_0 = 1$. If a is bigger than this value then $R_0 < 1$, and the virus cannot sustain an infection. (iii) Between these two extreme values, the total number of cells is less than λ/d. A minimum value is obtained for $a_{\min} = \sqrt{\lambda\beta k/u}$.

Therefore, if $a > a_{\min}$ then a CTL response will always increase the total number of cells. If, on the other hand, $a < a_{\min}$ then a weak CTL response may reduce the total number of cells, while a sufficiently strong response will increase it.

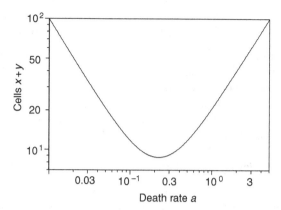

Fig. 6.1   The total abundance of cells, $x + y$, as a function of the death rate of infected cells, a, describes a curve with a minimum. For low and high values of a, the abundance of cells is close to the equilibrium level of cells in the absence of an infection. The lowest total number of cells is obtained for a death rate, a_{min}. If $a < a_{min}$, then increasing the death rate of infected cells (by an immune response) reduces the overall abundance of cells. If $a > a_{min}$, then increasing the death rate of infected cells increases the abundance of cells. Parameters: $\lambda = 1, d = 0.01, a = 0.5, \beta = 0.005, k = 50, u = 5, p = 1, b = 0.05$.

6.1.1   Persistent infection or clearance

The basic reproductive ratio of the virus in presence of the immune response, R_I, provides a simple condition for persistent infection or clearance. If $R_I > 1$ the infection will persist. If $R_I < 1$ it will be cleared.

We can ask how fast CTL have to kill infected cells in order to eliminate the infection. We obtain the condition

$$a' > a(R_0 - 1). \tag{6.7}$$

For example, if the basic reproductive ratio of the virus without an immune response is $R_0 = 5$, then CTL have to kill 4-times faster than the natural death rate of the infected cell. In other words 80% of infected cells have to be eliminated by the CTL response.

We can also ask how fast CTL have to kill in order to reduce virus load by a factor F. We find

$$F \equiv \frac{y^*}{\hat{y}} = \frac{R_0 - 1}{R_I - 1}. \tag{6.8}$$

Suppose $R_0 = 5$. A value of $R_I = 2$ would lead to a 4-fold reduction in virus load. In this case the CTL have to kill at 1.5-times the natural death rate of the infected cells, that is $a' = 1.5a$, which is equivalent to 60% of the cells killed by CTL.

6.1.2 *Variation in CTL responsiveness leads to a negative correlation between virus load and the magnitude of the CTL response*

Suppose patients differ in the parameter c, the rate at which the CTL response is induced once virus is present. The greater the CTL responsiveness, c, the greater the CTL response, z, and the smaller the equilibrium abundance of virus. (Remember, $a' \equiv cp/b$.) Therefore, the model predicts a negative correlation between virus load and the abundance of CTL. Patients with a weak CTL response have a high virus load; patients with a strong CTL response have a low virus load (Fig. 6.2).

6.2 Other self-regulating immune responses

As before we consider the basic model of virus dynamics and assume that the immune response may affect the parameters that determine viral reproduction. CTL-mediated lysis may increase the death rate of infected cells, a. Chemokines or cytokines released by CD8 or CD4 positive T cells may reduce the infectivity parameter β and/or the rate at which infected cells produce virus, k. Finally, antibody-mediated immune responses may enhance the removal rate of free virions, u. The effect of any immune response mechanism can be described in terms of the basic model of virus dynamics as increasing a, increasing u, reducing β or reducing k. We find that all immune responses reduce the abundance of free virus and infected cells and increase the abundance of uninfected cells. Immune responses that enhance the parameter a may increase or reduce the total number of cells, while all other immune responses increase this quantity.

As before we can define a basic reproductive ratio in the presence of immune responses,

$$R_{\mathrm{I}} = \frac{\lambda \beta f_\beta k f_k}{d(a + a')(u + u')}. \tag{6.9}$$

Here f_β and f_k are factors (between 0 and 1) by which the infectivity parameter (β) and the rate of virion production (k) are being reduced by immune responses, while a' and u' describe the effect of immune responses on removing infected cells and free virions. If $R_{\mathrm{I}} < 1$ then the virus is eliminated by the combined action of the immune responses. If $R_{\mathrm{I}} > 1$ the virus can establish a persistent infection. Provided $R_{\mathrm{I}} \gg 1$ (that is the virus is not close to elimination), the equilibrium abundances of infected cells and free virus are given by

$$y^* \approx \frac{\lambda}{a + a'} \tag{6.10}$$

and

$$v^* \approx \frac{\lambda k f_k}{(a + a')(u + u')}. \tag{6.11}$$

Thus only a' has a strong effect on y^*, while f_k, a' and u' can have strong effects on v^*.

Finally, we note that in models with self-regulating immune responses there is always a negative correlation between virus load and the abundance of immune mediators (immune cells, antibodies or chemokines). Patients with weak anti-viral responses have high virus load, while patients with strong anti-viral responses have low virus load.

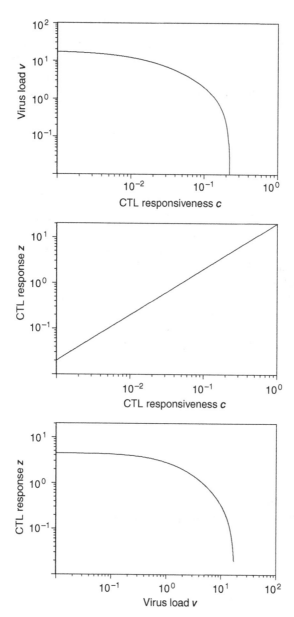

Fig. 6.2 The simplest model of an anti-viral CTL response, $\dot{z} = c - bz$, leads to a negative correlation between the equilibrium abundance of CTL, $\hat{z}$, and virus, $\hat{v}$, if patients vary in their CTL responsiveness, c. Patients with a strong CTL responsiveness have a high equilibrium abundance of CTL and a low virus load. Parameters: see Fig. 6.1 and $a = 0.5$.

6.3 A nonlinear CTL response: predator–prey dynamics

In this section, we consider a more dynamic and perhaps more realistic immune response against the virus. As before we study the effect of a CTL response that can eliminate infected cells, but this time the stimulation of the CTL response does not occur at a constant rate, but at a rate proportional to the abundances of CTL and infected cells, cyz. We obtain the following system of differential equations:

$$\begin{aligned}
\dot{x} &= \lambda - dx - \beta x v, \\
\dot{y} &= \beta x v - a y - p y z, \\
\dot{v} &= k y - u v, \\
\dot{z} &= c y z - b z.
\end{aligned} \tag{6.12}$$

The variable z denotes the magnitude of the CTL response. The rate of CTL proliferation is now given by cyz (rather than the constant c of Section 6.1). In the absence of stimulation, CTL decay at rate bz. Infected cells are killed by CTL at rate pyz. The parameter p specifies the rate at which CTL kill infected cells.

In this model, there is a minimum level of infected cells necessary to stimulate a CTL response. If $cy > b$ the CTL response will increase. The long-term outcome of the system depends on whether the equilibrium abundance of infected cells in the absence of a CTL response is above or below this threshold value. Without a CTL response (and provided that $R_0 > 1$) the system converges to the equilibrium

$$\begin{aligned}
x^* &= \frac{au}{\beta k}, \\
y^* &= \frac{\lambda}{a} - \frac{du}{\beta k}, \\
v^* &= \frac{\lambda k}{au} - \frac{d}{\beta}, \\
z^* &= 0.
\end{aligned} \tag{6.13}$$

If $cy^* < b$ this equilibrium is stable; the CTL response will not be activated.

If $cy^* > b$ the system shows damped oscillations to the equilibrium

$$\begin{aligned}
\hat{x} &= \frac{\lambda cu}{cdu + \beta bk}, \\
\hat{y} &= \frac{b}{c}, \\
\hat{v} &= \frac{bk}{cu}, \\
\hat{z} &= \frac{1}{p}\left(\frac{\lambda \beta ck}{cdu + \beta bk} - a\right).
\end{aligned} \tag{6.14}$$

There are two interesting remarks about this equilibrium. First, the equilibrium abundance of infected cells depends only on the immunological parameters b and c. Parameters determining the host cell dynamics only enter indirectly via the condition $cy^* > b$.

Second, the condition $cy^* > b$ is equivalent to the conditions $x^* < \hat{x}$, $y^* > \hat{y}$, and $v^* > \hat{v}$. Thus, in the above model, if there is an active CTL response, then it will reduce virus load and increase the equilibrium abundance of uninfected cells. But, as in the previous section, the total abundance of infected and uninfected cells, $\hat{x} + \hat{y}$, can be increased or decreased by a CTL response compared to $x^* + y^*$.

6.3.1 Virus load reduction

As in Section 6.1 we can define a basic reproductive ratio in the presence of the immune response as

$$R_\mathrm{I} = \frac{\lambda \beta k}{(a + p\hat{z})du}.$$ (6.15)

This is the number of infected cells that would arise out of any one infected cells, if the CTL response was at its equilibrium level, $\hat{z}$, and the uninfected cell population at its pre-infection level, $x_0 = \lambda/d$. This leads to

$$R_\mathrm{I} = 1 + \frac{\beta bk}{cdu}.$$ (6.16)

Note that R_I is always greater than one, therefore elimination of the virus is not possible. More precisely, the equilibrium abundance of the virus is always positive. (Of course the derivation of 6.16 pre-supposes $cy^* > b$ and hence $R_0 > 1$.)

But as before we can ask how fast CTL have to kill in order to reduce virus load by a factor F. Perhaps surprisingly, we find exactly the same relation:

$$F := \frac{y^*}{\hat{y}} = \frac{R_0 - 1}{R_\mathrm{I} - 1}.$$ (6.17)

As before if, for example, $R_0 = 5$, then a 4-fold reduction of virus load requires $R_\mathrm{I} = 2$ and hence $p\hat{z} = 1.5a$. In other words, 60% of infected cells have to be killed by the CTL response.

6.3.2 Variation in immune responsiveness

Let us now study the relation between CTL responsiveness, c, CTL response, $\hat{z}$, and virus load, $\hat{y}$ or $\hat{v}$. We shall see that CTL responsiveness determines virus load, but is not necessarily reflected by the magnitude of the CTL response at equilibrium.

Suppose patients differ in their CTL responsiveness, c, while all other parameters are constant. Figure 6.3 shows virus load, v, and the abundance of CTL, z, as functions of c. Virus load is a declining function of c. The abundance of CTL is an increasing function of c, which saturates for high levels of c. Thus at high levels of c, variation in c has a large effect on virus load, but only a small effect of the magnitude of the CTL response. The correlation between the CTL response and the virus load is negative, but there can be large variation in virus load with only small changes in the abundance of CTL.

Let us now assume that the parameters c and p are not independent of each other. For instance, the rate at which CTL get stimulated to divide following encounter with

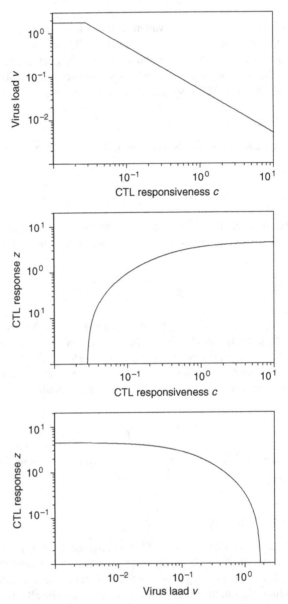

Fig. 6.3 The nonlinear model of the CTL response also leads to a negative correlation between CTL abundance and virus load if patients differ in their CTL responsiveness, c. While virus load, v, is a declining function of c, the CTL response, z saturates for high values of c. Thus, the negative correlation can be weak depending on the range of variation of c. Parameters: see Fig. 6.2.

infected cells may be linked to the rate at which they can eliminate infected cells. A good response may imply both a high rate of CTL proliferation and a high efficacy of killing infected cells. Figure 6.4 shows virus load, v, and CTL abundance, z, as functions of c under the assumption that c is directly proportional to p. As before we find that virus load is a declining function of c. A high CTL responsiveness implies a low virus load. The functional dependence of z on c is, however, more complex: there is a one-humped function. Initially z increases, reaches a maximum, and then declines again for high values of c. The reason for this behaviour is that the equilibrium abundance of z depends on both c $(=p)$ and the equilibrium abundance of y. For very low values of c there is a high virus load, but only a very inefficient stimulation of the CTL response. For high values of c there is efficient stimulation of the CTL response, but the equilibrium abundance of infected cells is very low.

The fact that the CTL abundance, z, declines for high values of c generates (over a large range of variation of v) a *positive* correlation between virus load and CTL abundance (Fig. 6.4). Patients with high CTL responsiveness have a low virus load and a low abundance of CTL. Patients with lower CTL responsiveness have a higher virus load and a higher abundance of CTL. For very low values of c the correlation becomes negative again: patients with very small values of c have almost no CTL response and very high virus load.

In such a scenario, cross-sectional studies comparing virus load and CTL abundance in a large number of different patients can either lead to a positive correlation, a negative correlation or no correlation at all depending on the range of variation of c. Interestingly a negative correlation (or no correlation) implies only a weak effect of CTL on virus load reduction. A positive correlation, on the other hand, occurs for the parameter range where CTL have a strong effect on virus load reduction!

The model can be extended to include immune responses other than CTL-mediated lysis. Similar results will be obtained as long as the immune stimulation term is proportional to the product of antigen abundance and abundance of immune cells. This underlines the generality of the idea that immune responsiveness determines virus load at equilibrium but there need not be a simple correlation between immune responsiveness and the magnitude of the immune response in comparisons among different infected individuals. Virus load is a better indicator of immune responsiveness than the abundance of immune cells.

6.4   A linear immune response

Next we study a third model where the CTL response is induced at a rate proportional to the abundance of infected cells, but independent of the abundance of CTL:

$$\dot{x} = \lambda - dx - \beta xv,$$
$$\dot{y} = \beta xv - ay - pyz,$$
$$\dot{v} = ky - uv,$$
$$\dot{z} = cy - bz.$$

(6.18)

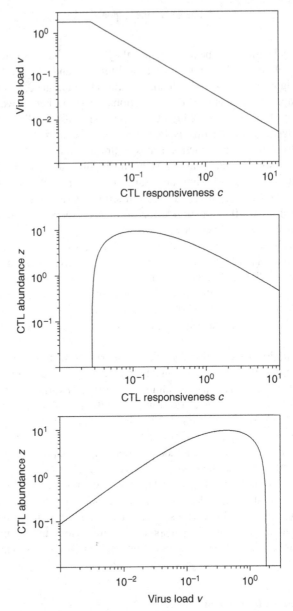

Fig. 6.4 A more complicated picture is obtained if we assume that variation in c also implies variation in p, which means that the rate of proliferation of CTL when encountering a target cell is linked to the rate at which CTL kill the target cell. In this case, there is a one-humped function for the equilibrium abundance of CTL versus virus load. Depending on the range of variation of c (and p), the correlation can be positive or negative. For high levels of c, however, we find a *positive* correlation. That is, if the immune response is effective in controlling the virus then patients with higher CTL responsiveness will have lower virus loads and lower equilibrium abundance of CTL. Parameters: see Fig. 6.2, but $p = c$.

There is no activation threshold for the immune response. Provided $R_0 > 1$ the system converges to the equilibrium

$$\hat{x} = \frac{\beta' ab - cdp + \sqrt{(\beta' ab - cdp)^2 + 4\beta'^2 \lambda bcp}}{2\beta'^2 b},$$

$$\hat{y} = b\hat{z}/c, \tag{6.19}$$

$$\hat{v} = k\hat{y}/u,$$

$$\hat{z} = (\beta'\hat{x} - a)/p.$$

If the parameter c varies while all other parameters remain constant, we find, as before, that virus load is a decreasing function of c, while CTL abundance is an increasing function of c. The correlation between virus load and CTL abundance is negative. For the situation where CTL responsiveness, c, is directly linked to the rate of CTL-mediated lysis, p, we find as in the previous section that the equilibrium abundance of CTL is given by a one-humped function of c and that the correlation between virus load and CTL abundance is positive (in the parameter region where CTL have a strong effect on reducing virus load). Thus the less convenient model (6.18) leads to the same insights as the previous model. The main difference is that the equilibrium virus load of model (6.18) does not only depend on the immunological parameters, which is more realistic. The model is also less oscillatory.

6.5 Dynamic elimination

The models in Sections 6.3 and 6.4 always admit an equilibrium with a positive virus load; the immune response can in the long term not eliminate the virus. Immediately after infection, however, there can be oscillations with large amplitude. Virus load increases at first dramatically and subsequently falls to very low levels. The minimum virus load of this first oscillation may correspond to a virus population size of only a few cells or even less than one cell. Thus the virus population may be eliminated by the magnitude of the first oscillation, although there would be a stable equilibrium corresponding to a persistent infection. We propose to call this phenomenon 'dynamic elimination'. For a given virus infection the chances of dynamic elimination increase with the quality of the immune response. A higher immune responsiveness, c, makes it more likely that the virus population is eliminated after the initial peak. A low immune responsiveness, on the other hand, favors a persistent infection. In this context, Dominik Wodarz proposed the novel idea that a long-lived CTL response (that is, a *CTL memory response*) might be required for eliminating a virus infection or reducing virus load to very low levels. He also showed that the probability of dynamic elimination increases with the 'memory quality' of the CTL response. (For more details we refer to Wodarz *et al.* 2000*b*.)

6.6 The simplest models of immune response dynamics

In the previous sections we have studied how immune responses reduce virus load in the basic model of virus dynamics that describes uninfected cells, infected cells and free virus. The virus population is held in check by a combination of target cell limitation and immune responses. In this section, we take a simplified approach: we neglect target cell control and assume that the virus population is only controlled by immune responses.

Consider the simplest such model:

$$\dot{v} = v(r - pz),$$
$$\dot{z} = c - bz. \tag{6.20}$$

Here the variable v describes the virus population (which can be seen either as abundance of free virions or infected cells), whereas the variable z denotes the abundance of immune cells (or antibodies) against the virus. In the absence of immunity, the virus grows exponentially at the rate, rv. Immune responses remove virus at the rate, pvz. Once antigen is encountered, the immune cells expand at a constant rate, c, and die at the rate bz.

The immune response will converge toward the equilibrium level $z^* = c/b$. There are two possibilities for what happens to the virus population: (i) if $r < pc/b$ then the virus is controlled by the immune response, and viral abundance will decline to zero; (ii) if $r > pc/b$ then the immune response is unable to control the virus, which will grow exponentially (and kill the patient). Case (i) may correspond to an acute viral infection with subsequent clearance, while case (ii) corresponds to an uncontrolled acute infection with rapid progression to very high virus load.

The second model assumes a linear term for immune stimulation: immune cell abundance increases in response to antigen at the rate cv. This leads to

$$\dot{v} = v(r - pz),$$
$$\dot{z} = cv - bz. \tag{6.21}$$

The system converges in damped oscillations to the stable equilibrium

$$v^* = \frac{rb}{cp}, \quad z^* = \frac{r}{p}. \tag{6.22}$$

The dynamics of the system can be interpreted as an initial phase of infection with high virus load and subsequent down-regulation of the virus to a lower equilibrium virus load, corresponding to a persistent infection.

As a third model, we assume that immune cells grow at a rate proportional to the abundance of virus times the abundance of immune cells. In other words, their rate of cell division is a linear function of the abundance of virus. We obtain

$$\dot{v} = v(r - pz),$$
$$\dot{z} = z(cv - b). \tag{6.23}$$

The system shows undamped oscillations around the neutrally stable equilibrium

$$v^* = \frac{b}{c}, \quad z^* = \frac{r}{p}. \tag{6.24}$$

Such dynamics are unrealistic, because virus and immune cells remain in unabated oscillations; the model would require further additions for damping the oscillations. This third model is in fact the simplest Lotka–Volterra equation describing predator–prey dynamics. The virus is the prey, the immune system the predator.

It is of interest to study a model, which is essentially a combination of models 1, 2 and 3:

$$\dot{v} = v(r - pz),$$
$$\dot{z} = c_1 + c_2 v + c_3 vz - bz. \tag{6.25}$$

Here the stimulation of the immune response is described by three terms, a constant production rate, c_1, a production term proportional to virus abundance, $c_2 v$, and a production rate proportional to the product of virus times immune cell abundance, $c_3 vz$. If $r > pc_1/b$, the system settles in damped oscillations to the equilibrium

$$v^* = \frac{rb - c_1 p}{c_2 p + c_3 r}, \quad z^* = \frac{r}{p}. \tag{6.26}$$

If $r < pc_1/b$ the virus infection is cleared. Thus the model gives the possibility both of a transient and of a persistent virus infection.

A main characteristic of HIV is to impair immune responses. We may describe this feature of HIV by a term, $-uvz$, in the equation for the immune response. The parameter u describes the rate at which the virus can eliminate immune cells. We then have

$$\dot{v} = v(r - pz),$$
$$\dot{z} = c_1 + c_2 v + c_3 vz - bz - uvz. \tag{6.27}$$

There are four possibilities; two as for eqn (6.25) and an additional two which derive, in effect, from the possibility that the 'c_3' parameter in eqn (6.25) is now '$c_3 - u$', which can be negative.

(i) If $r > pc_1/b$ and $c_2 p > (u - c_3)r$ then the system converges in damped oscillations to the stable equilibrium:

$$v^* = \frac{rb - c_1 p}{c_2 p + (c_3 - u)r}, \quad z^* = \frac{r}{p}. \tag{6.28}$$

(ii) If $r < pc_1/b$ and $c_2 p > (u - c_3)r$ then the virus is cleared by the immune response.

(iii) If $r > pc_1/b$ and $c_2 p < (u - c_3)r$ then the virus population cannot be controlled by the immune response; there is no stable equilibrium, the virus grows unboundedly.

(iv) If $r < pc_1/b$ and $c_2 p < (u - c_3)r$ then for low virus load, the virus is eliminated, whereas for high virus load it completely escapes from immune control.

These results can be established by a simple phase plane analysis.

6.6.1   Variation in immune responsiveness

In all of the above models, whenever there is a stable equilibrium between the virus and the immune response, the equilibrium abundance of the immune response does not depend on the immune responsiveness parameter c. Instead the immune response grows to whatever level is required to control the virus population; c does not alter this level, only the rate at which it is reached.

Thus if a collection of patients only vary in c, then all will have the same equilibrium abundance of the immune response in a persistent infection. Of course, patients with a high c have a low equilibrium virus load. In addition, variation in c can determine whether a persistent infection is at all possible or whether the virus infection is cleared by the immune response.

As noted above, however, variation in c is likely to be accompanied by variation in p. The two parameters are biologically linked. An efficient immune response is characterized by both a high c and a high p. High values of p lead to a low equilibrium abundance of the immune response. High values of p and c lead to a low equilibrium virus load. Thus there is a positive correlation between virus load and immune response if patients are compared cross-sectionally. In a good responder both v^* and z^* are low; in a weak responder both quantities are high.

6.7   Experimental observations: HTLV-1 and HIV-1, 2

The human T-cell leukaemia virus (HTLV-1) was the first known human retrovirus. It was discovered by Robert Gallo's group at the NIH in the early 1980s. It has a similar genetic organisation as HIV-1 and also infects CD4 positive T-cells. Unlike HIV-1, however, it is not a lentivirus, but an oncovirus. It can cause cancer by transforming infected cells. HTLV-1 is also associated with an inflammatory disease of the central nervous system called HTLV-1 associated myelopathy or tropical spastic paraparesis (HAM/TSP). Charles Bangham and his colleagues at the Imperial College London observed in healthy HTLV-1 carriers a strong *positive* correlation between the abundance of HTLV-1 and the abundance of specific CTLs. In HAM/TSP patients, on the other hand, they found no such correlation. They also observed that HAM/TSP patients have on average a significantly higher virus load that healthy carriers. Finally, Bangham and colleagues demonstrated that in a large cohort of Japanese patients a particular HLA-genotype is associated with a lower HTLV-1 provirus load and protection from HAM/TSP. This finding was in fact the first demonstration of a protective effect of a particular HLA type in a human infection.

In HIV-1, Graham Ogg working with Andrew McMichael's group at Oxford and David Ho's group at the Aaron Diamond AIDS Research Center in New York found a strong *negative* correlation between viral load and anti-viral CTL. This study is also in agreement with similar findings in HIV-2 infection obtained by Hilton Whittle in The Gambia.

Our mathematical models can suggest explanations for these observations. In asymptomatic carriers of HTLV-1 the virus is controlled by an efficient CTL response. As outlined in Section 6.3.2, in this case we expect a *positive* correlation between virus and CTL abundance. This is what is observed. In HAM/TSP patients there is, in general, no efficient CTL response and there is also no correlation between virus and CTL abundance. The mathematical model in Section 6.3 predicts no correlation (or a weak negative correlation) for the parameter range where CTL have little effect on reducing virus load. For HIV infection, Dominik Wodarz extended the model of Section 6.3 to include the fact that HIV can impair immune responses. In this case, the theoretical prediction is a *negative* correlation between abundance of virus and CTL, as observed by Ogg and Whittle.

6.8   Further reading

Virus load is strongly correlated with the risk of disease in persistent virus infections such as HIV-1 (Tersmette *et al.* 1988; Asjo *et al.* 1990; Munoz *et al.* 1992; Connor *et al.* 1993; Piatak *et al.* 1993; Mellors *et al.* 1995; Haynes *et al.* 1996; Little *et al.* 1999; Phillips *et al.* 1999), HTLV-I (Kira *et al.* 1991, 1992; Bangham 1993; Nagai *et al.* 1998), hepatitis B virus (Burk *et al.* 1994; Fong *et al.* 1994) and hepatitis C virus (Fabrizi *et al.* 1998; Neumann *et al.* 1998).

Cytotoxic T cells play a critical part in limiting viral replication (see Lin and Askonas 1981; Koenig *et al.* 1993; Jacobson *et al.* 1990; Bangham *et al.* 1996, 1999; Daenke *et al.* 1996; Kagi *et al.* 1996; McMichael *et al.* 1995; Zinkernagel and Hengartner 1994, 1997; Zinkernagel 1996). Jeffery *et al.* (1999) describe the effect of HLA alleles in HTLV-1 infection. Ogg *et al.* (1998) describe a negative correlation between viral load and specific CTL frequency among patients with HIV-1 infection. There is also evidence that CTL may be able to significantly reduce virus load and delay progression to AIDS (Schmitz *et al.* 1999).

For mathematical modelling of immunological processes we refer to Bell (1970*a,b*, 1971*a,b*), Anderson and Bell (1971), Perelson *et al.* (1976, 1978, 1980), Perelson and Oster (1979), Perelson and Bell (1982), Agur (1989), Agur *et al.* (1989, 1991), Anderson *et al.* (1989), Segel and Perelson (1989), McLean and Kirkwood (1990), Marchuk *et al.* (1991*a,b*), Celada and Seiden (1992, 1996), De Boer *et al.* (1992), McLean and Nowak (1992), Nelson and Perelson (1992), Perelson and Weisbuch (1992), Seiden and Celada (1992), Kepler and Perelson (1993), McLean (1993), Perelson *et al.* (1993), Antia and Koella (1994), De Boer and Boerlijst (1994), Essunger and Perelson (1994), Gupta *et al.* (1994), McLean (1994), Segel and Jager (1994), De Boer and Perelson (1995), Fishman and Perelson (1995), McLean and Mitchie (1995), Borghans *et al.* (1996), De Boer and Boucher (1996), Haase *et al.* (1996), Hetzl and Anderson (1996), Mehr *et al.* (1996), Krakauer and Payne (1997), Levin *et al.* (1997), Marchuk (1997), McLean *et al.* (1997), Perelson *et al.* (1997*a,b*), De Boer *et al.* (1998), De Boer and Noest (1998), De Boer and Perelson (1998), Kirschner *et al.* (1998), Mittler *et al.* (1998), Mohri *et al.* (1998), Segel (1998), Wodarz and Nowak (1998*a–d*), Grossman *et al.* (1999), Gupta and Anderson

(1999), Hlavacek *et al.* (1999), Jin *et al.* (1999), Krakauer and Nowak (1999), Segel and Bar-Or (1999), Wolthers *et al.* (1999) and Ramratnam *et al.* (2000). Nowak and Bangham (1996) develop some of the models discussed in this chapter and describe the effect of immune responsiveness on virus load. Wodarz *et al.* (2000*a*) analyse the correlations between CTL abundance and virus load in HIV and HTLV infection. Wodarz *et al.* (2000*b*) suggest that the role of CTL memory is to eliminate virus infections.

7

HOW FAST DO IMMUNE RESPONSES ELIMINATE
INFECTED CELLS?

In order to understand the role of the immune system in limiting HIV-1 replication, it is critical to know to what extent the rapid turnover of productively infected cells is caused by viral cytopathicity or by immune-mediated lysis. In this chapter, we show data that many HIV-1 infected patients contain cytotoxic T lymphocytes (CTL) that lyse target cells *in vitro*—at plausible CTL to target ratios—with half-lives of less than 1 day. In 23 patients with CD4 counts ranging from 10 to 900 per μl, the average rate of CTL-mediated lysis corresponds to a target cell half-life of 0.7 days. We develop mathematical models to calculate the turnover rate of infected cells subjected to immune-mediated lysis and viral cytopathicity and to estimate the fraction of cells which are killed by CTL as opposed to virus. The models provide new interpretations of drug treatment dynamics and explain the otherwise puzzling fact that the observed rate of virus decline is roughly constant for different patients. We conclude that in HIV-1 infection CTL-mediated lysis can reduce virus load by limiting virus production, with small effects on the half-life of infected cells.

In Chapter 4 we have seen that the initial rate of virus decline during anti-HIV drug treatment is approximately 30% per day. This has been interpreted that productively infected cells have a half-life of about 2 days. There is little variation in this half-life, ranging from about 1 to 4 days in patients with CD4 cell counts between 20 and 500 (Fig. 7.1). There is no correlation between the rate of viral turnover and CD4 cell count or virus load. The important question is whether the observed turnover rate of infected cells is caused by viral cytopathicity or immune-mediated clearance mechanisms. If we postulate that CTL-mediated lysis determines the life-span of a productively infected cell, then it is surprising to find so little variation in the observed half-life of infected cells and no correlation with the disease stage (CD4 cell count) of a patient. On the other hand, if cell death is only due to viral cytopathicity then immune-mediated lysis may have no effect on reducing virus production.

In Section 7.1 we present data from various studies of CTL-mediated lysis. We calculate the rate at which target cells are eliminated by CTL *in vitro*. The rate depends on the ratio of effector to target cells, and therefore it is important to adjust this ratio to what one would expect to occur *in vivo*. Furthermore, it is important to use assays of fresh cytotoxicity where CTL are directly isolated from a patient and not subjected to *in vitro* activation. Nevertheless, it is still impossible to know whether the *in vitro* measurements reflect the *in vivo* clearance rates, but it is interesting to note that the *in vitro* clearance rates are in reasonable agreement with estimates from drug treatment

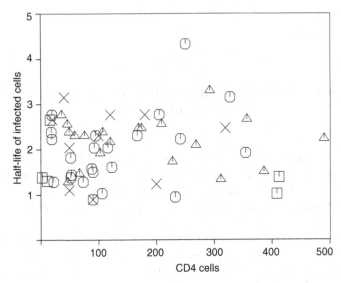

Fig. 7.1 The turnover rate of productively infected cells from 59 HIV-1 infected patients treated with anti-viral drugs. The average death rate of productively infected cells, calculated from the exponential decline of plasma virus load following treatment, is 0.37 ± 0.18 per day, corresponding to a half-life of 1.9 days. The minimum half-life is about 1 day. There is no correlation between infected cell half-life and CD4 cell count. Symbols represent patients from different studies: triangles (Ho *et al.* 1995), circles (Wei *et al.* 1995), ×-s (Nowak *et al.* 1995), squares (Perelson *et al.* 1996).

studies. Drug treatment studies suggest *in vivo* half-lives of productively infected cells of about 1–3 days. The *in vitro* clearance rate of the CTL response of many patients is between 0.5 and 5 days. (There are also patients where a fresh CTL response is undetectable.)

In Section 7.2 we present detailed mathematical models to quantify the effect of CTL killing on infected cell half-life and virus production. In contrast to the basic models of virus dynamics and immune clearance (as discussed in Chapters 3–5), we explicitly consider the life-cycle of a virus-infected cell: immediately after infection, a cell may not be visible to CTL-mediated lysis; after some time enough viral proteins have been produced and the cell becomes a target for CTL recognition; again after some time the cell will release new virions. Finally the cell may die due to viral cytopathicity. We will calculate the fraction of cells eliminated by CTL (as opposed to viral cytopathicity). We will calculate the average life-time of a cell and the average amount of virions released by an infected cell. We will calculate the total amount of virion production inhibited by CTL-mediated lysis. We will show that for a cytopathic virus CTL-mediated lysis can have a great effect on reducing the overall amount of virion production without greatly reducing the life-time of an infected cell.

In Section 7.3 the models will be used to analyse virus decay slopes during treatment. For a cytopathic virus it is possible to obtain very similar decay slopes for patients with very different CTL responses. For a non-cytopathic virus the observed decay slope depends on the efficacy of the CTL response. Section 7.4 compares the HIV-1 data to data obtained for hepatitis B virus (HBV), as discussed earlier. The data suggest that HIV-1 is cytopathic and HBV is non-cytopathic *in vivo*.

7.1   The rate of CTL-mediated lysis *in vitro*

In HIV-1 infected patients, anti-viral CTL arise early in infection and are present within uncultured circulating PBMC as a subset of CD8+ cells. The rate of CTL-mediated killing should influence both the half-life of virus-infected cells and the amount of virus production. Figure 7.2 shows the *in vitro* half-life of target cells from assays of specific cytotoxicity by fresh PBMC. Figure 7.2A shows time-resolved decay curves of target cells in the presence of PBMC from an HIV-1 positive patient (who has been infected for about 5 years and has currently a CD4 cell count of 540) at various PBMC to target ratios. From the slope of the decay curve we can calculate the death rate of target cells. At a PBMC to target ratio of 64 : 1 the death rate of target cells corresponds to a half-life of 12 h. Figure 7.2B shows the half-lives of target cells for different PBMC to target ratios.

Clearly, the above data are results from *in vitro* measurements and it is uncertain how accurately they reflect the rate of CTL-mediated lysis *in vivo*. We argue, however, that using uncultured PBMC and adding a small amount of target cells may provide conditions that are as close as possible to the *in vivo* situation. The fraction of cells infected with HIV-1 and expressing HIV-1 RNA in lymph nodes can be as high as 3–6% in asymptomatic subjects, although this may vary depending on the lymphoid micro-environment. This implies an overall ratio of lymphocytes to infected target lymphocytes of at least 15–30 : 1. In patients with lower viral burden the ratio will be higher. Therefore, by using PBMC to target ratios of 25 : 1 or larger we may simulate the *in vivo* situation.

The above experiments were carried out with peptide pulsed B cell lines as targets. Figure 7.3 shows the death rate of HIV-1 infected CD4 cells in the presence of uncultured PBMC from an HIV-1 positive patient (an asymptomatic individual who has been infected for at least 3 years and has a CD4 cell count of 440 per μl). At a PBMC to target ratio of 64 : 1 we find 17% specific lysis after 8 h corresponding to a target cell half-life of 1.2 days. At a ratio of 32 : 1 we obtain a half-life of 2.4 days.

Results derived from published studies of fresh responses by many different groups can also be used to calculate target cell half-lives (Fig. 7.4). In some patients, the half-life of target cells is less than 12 h at PBMC to target ratios of 50 : 1 or greater. On the other hand many HIV-1 infected individuals do not show strong responses and target cell half-life due to CTL killing would exceed 48 h.

To explore a possible correlation between the rate of CTL-mediated lysis and disease stage of a patient, we calculated the half-life of target cells in the presence of fresh PBMC from 23 patients with CD4 cell counts ranging from 10 to 900. CTL activity was determined against targets expressing five different HIV proteins: Gag, Pol, Env, Nef or

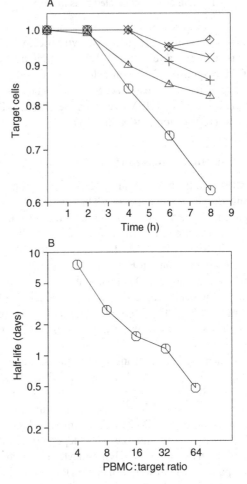

Fig. 7.2 The average life-time of target cells can be calculated from assays of specific cytotoxicity. (A) Time-resolved decay curve of peptide sensitized target cells from HIV-1 infected asymptomatic patient. The assays of fresh killing were performed using PBMC separated directly. Targets were HLA B8-matched BCL, untreated or pre-pulsed with 1 μM p17-3 (GGKKKYKL), an epitope in p17 to which this patient makes a predominant response. The decay rate of target cells can be directly obtained from the slope of the decay curves. There is an initial 2 h time delay where no killing can be observed. This may be a consequence of the effector cells needing time to attach to and kill the first targets. Between 2 and 8 h there is a roughly constant killing rate. PBMC : target ratios are 64 : 1 (circle), 32 : 1 (triangle), 16 : 1 (+), 8 : 1 (×), and 4 : 1 (diamond). Different amounts of PBMC are added to the same number of targets. At a ratio of 64 : 1 target cell half-life is 11.6 h (in the first 8 h) and 8.7 h (if calculated from the slope between 2 and 8 h). (B) Half-life of target cells (calculated from 0–8 h) versus PBMC : target ratio gives roughly a straight line in a double logarithmic plot with a slope of about −1. This suggests that the killing rate of target cells is simply proportional to the number of effector cells in the assay.

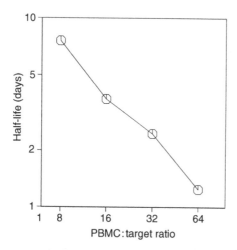

Fig. 7.3 Half-life of HIV infected target cells subjected to uncultured CTL-mediated lysis. Effector cells were freshly isolated PBMC from an HLA B8-positive HIV-1 infected patient. Targets were B8-matched C8166 cells used 48 h after infection with 10 $TCID_{50}$ of HIV IIIB.

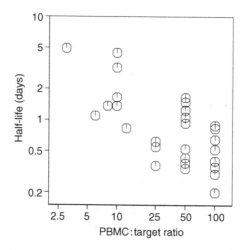

Fig. 7.4 Half-lives of target cells derived from published studies of lysis by fresh PBMC from HIV-1 infected patients. Points represent individual patients.

Tat. Figure 7.5 shows the calculated half-life of target cells with respect to the strongest response among the Gag, Pol, Nef or Tat specific CTL in each patient. At a PBMC to target ratio of 50 : 1 the average rate of CTL-mediated lysis is 1.0 ± 0.57 per day which corresponds to a half-life of about 0.7 days. There is no correlation between the CD4

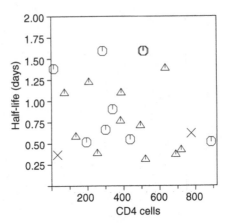

Fig. 7.5 Half-life of target cells in assays of fresh cytotoxicity versus CD4 cell count of 23 patients. For each patient we show the maximum response among the anti-Gag (circle), anti-Pol (triangle), or anti-Nef ($\times$), CTL at PBMC : target ratios of 50 : 1. (The anti-Tat response was never the maximum response in any of the patients.) The average rate of CTL-mediated lysis is 1.03 ± 0.57 per day which corresponds to a half-life of 0.68 days. There is no correlation between CTL-mediated lysis and CD4 cell count. Data are from Klenerman *et al.* (1996*a*).

count of a patient and the half-life of target cells. Note that the lack of correlation between viral turnover and CD4 count (Fig. 7.1) is paralleled by a lack of correlation between the fresh CTL response and CD4 count (Fig. 7.5).

The calculated average rate of anti-Env-mediated lysis *in vitro* is 1.24 ± 0.66 per day corresponding to a half-life of about 0.6 days. In contrast to the anti-Gag, -Pol, -Nef or -Tat responses, however, which are mostly due to CTL activity, the anti-Env response might contain a large proportion of antibody dependent cell cytotoxicity (ADCC).

7.2   The dynamics of CTL-mediated lysis

Given these findings, we will now explore whether CTL-mediated lysis can make a significant contribution to the turnover rate of HIV-infected cells *in vivo*. If HIV directly kills productively infected cells, does CTL-mediated destruction play a role in the overall decay rate of infected cells? We develop a mathematical model to relate the expected life-time of an infected cell to the rate of CTL-mediated clearance mechanisms and virus cytopathicity and to calculate the amount of virus production inhibited by CTL. Consider a model with three compartments of infected cells representing different stages of the virus life-cycle: y_1 are newly infected cells, which are not yet producing free virus and are not targets for CTL killing; y_2 are cells which are still not producing virus but can be killed by CTL; whereas y_3 are cells which are both producing free virus and can be killed by CTL.

In the following we will explore two specific cases. In Model 1 we will assume that the transitions between stages occur at fixed time points, whereas in Model 2, we assume that they occur at constant rates. Models 1 and 2 are limiting cases on opposite ends of the spectrum; the reality is likely to be found in between.

7.2.1 Model 1

In Model 1, we assume that the transitions from y_1 to y_2 and from y_2 to y_3 occur at fixed times t_1 and t_2 in the life-cycle of an infected cell. This means a cell gets invaded by virus at time $t = 0$ and at time t_1 enough new viral proteins have been produced such that the cell becomes a potential target for CTL-mediated killing. The death rate of a target cell due to CTL is given by α. At time t_2 the cell starts to produce free virus particles. This increases the death rate by an amount c, which is due to virus cytopathicity. Thus, we have a stepwise model for cellular decay. Between time 0 and t_1 the death rate is 0, between times t_1 and t_2 the death rate is α, after time t_2 the death rate is $\alpha + c$.

These assumptions lead to an average life-time of infected cells of

$$T = t_1 + \frac{1}{\alpha} - \frac{c}{\alpha(\alpha + c)} e^{-\alpha(t_2 - t_1)}. \tag{7.1}$$

In the absence of CTL-mediated lysis ($\alpha = 0$), the average life-time is $T = t_2 + (1/c)$. The half-life of an infected cell is

$$T_{1/2} = t_1 + \frac{1}{\alpha} \log 2 \quad \text{if } t_2 > t_1 + \frac{1}{\alpha} \log 2,$$

$$T_{1/2} = t_2 + \frac{1}{\alpha + c} [\log 2 - \alpha(t_2 - t_1)] \quad \text{if } t_2 < t_1 + \frac{1}{\alpha} \log 2. \tag{7.2}$$

Note that the relation between half-life and average life-time is not simply $T_{1/2} = T \log 2$, because infected cell death is a combination of different exponential declines. The average duration of a cell producing new virus is

$$T_v = \frac{1}{\alpha + c} e^{-\alpha(t_2 - t_1)}. \tag{7.3}$$

In the absence of CTL this increases to $T_0 = 1/c$. The fraction of cells which are killed by CTL before virus production sets in is

$$f = 1 - e^{-\alpha(t_2 - t_1)}. \tag{7.4}$$

The total fraction of cells killed by CTL as opposed to viral cytopathicity is

$$F = 1 - \frac{c}{\alpha + c} e^{-\alpha(t_2 - t_1)}. \tag{7.5}$$

Assuming constant viral production after time t_2, the fraction of virus production inhibited by CTL-mediated lysis can be defined as $1 - T_v/T_0$, which is equivalent to F.

Table 7.1 *Effect of CTL-mediated lysis on the half-life of infected cells, $T_{1/2}$, and the fraction, F, of cells killed by CTL according to Model 1*

Rate of CTL killing (per day)	Cytopathic virus early target formation		Cytopathic virus late target formation		Non-cytopathic virus	
	$T_{1/2}$ (days)	F	$T_{1/2}$ (days)	F	$T_{1/2}$ (days)	F
0 (0%)	2.7	0	2.7	0	70	0
0.12 (2%)	2.2	0.29	2.3	0.24	6.2	0.92
0.31 (5%)	1.7	0.54	1.9	0.45	3.1	0.97
0.63 (10%)	1.2	0.75	1.6	0.64	2	0.99
1.34 (20%)	0.8	0.91	1.3	0.8	1.4	0.99
3.06 (40%)	0.5	0.99	1.1	0.92	1.1	0.998

We assume that virus production starts $t_2 = 1$ day after infection of a cell. For the cytopathic virus, the death rate of virus-producing cells is $c = 0.4$ per day, for the non-cytopathic virus this death rate is $c = 0.01$ per day. For the cytopathic virus, we explore early target formation ($t_1 = 0.3$ days) and late target formation ($t_1 = 0.9$ days). For the non-cytopathic virus, the value of t_1 makes almost no difference and we choose $t_1 = 0.9$. The rate of CTL killing, α, is given per day (in brackets the equivalent percentage specific lysis in a 4 h assay). The results show that for a cytopathic virus differences in the rate of CTL-mediated lysis have only small effects on the half-life of infected cells (ranging from 0.5 to 2.7 days), whereas for the non-cytopathic virus there is large variation in infected cell half-life (ranging from 1.1 to 70 days). In both cases CTL-mediated lysis can kill a large fraction of infected cells and therefore significantly reduce virus production.

Table 7.1 gives numerical values for the effect of CTL killing. For a cytopathic virus the half-life of infected cells does not vary much in the presence of weak or strong CTL responses, but the fraction of infected cells killed by CTL and the amount of virus production inhibited by CTL can be greatly affected by the rate of CTL-mediated lysis. For example, a response equivalent to 10% lysis in 4 h can eliminate about 70% of infected cells; the remaining 30% are eliminated by virus cytopathic effects. For a non-cytopathic virus, differences in CTL activity lead to large variation in infected cell half-life. Responses equivalent to 10% lysis in 4 h eliminate 99% of infected cells.

7.2.2 Model 2

In Model 2, we assume that y_1 turns into y_2 at a constant rate, a; y_2 turns into y_3 at rate b; y_3 cells are killed by virus at rate c, and y_2 and y_3 cells are killed by CTL at rate α. In this model, the average life-time of a cell is

$$T = \frac{1}{a} + \frac{1}{\alpha + b} + \frac{b}{(\alpha + b)(\alpha + c)} \tag{7.6}$$

and the average duration of a cell producing virus is

$$T_v = \frac{b}{(\alpha + b)(\alpha + c)}.$$ (7.7)

In the absence of a CTL response, $\alpha = 0$, this duration increases to $T_0 = 1/c$. The fraction of cells killed by CTL is

$$F = 1 - \frac{bc}{(\alpha + b)(\alpha + c)},$$ (7.8)

which is again equivalent to the total amount of virus production inhibited by CTL.

Between the two limiting cases, described by Models 1 and 2, lies a spectrum of models where the transition from y_1 to y_2 (and from y_2 to y_3) is not given by a simple one step process with an exponential distribution and also does not occur after a fixed length of time. We considered a Poisson process, where a number, n, of events have to accumulate (at certain rates) before the transition occurs. If $n = 1$ we obtain Model 1, and if n is very large we obtain Model 2. For intermediate values of n, the resulting decay is not strictly exponential but for practical purposes may be indistinguishable from an exponential decay. In short, reality is likely to be bracketed by the Models 1 and 2.

7.3   Virus decay slopes

The above models also provide new insights into the dynamics of virus decline following drug treatment. Before drug treatment the virus population is at steady-state and infected cells, y_1, are produced at a constant rate, β (which depends on the abundance of infectable cells, virus load, cytokine levels, etc). The effect of drug treatment is to reduce β to (almost) 0. Reverse transcriptase inhibitors (RTIs) prevent the infection of new cells, while protease inhibitors (PIs) render virus particles, which are released from already infected cells, non-infectious. Therefore, RTIs bring β to (almost) 0 as soon as they achieve a high enough concentration within the patient, while PIs allow for a short time infection of new cells by virus particles which have been produced before the drug was given. If the free virus half-life is short, then the difference is negligible.

In both Models 1 and 2, this leads to an (approximately) exponential decline in plasma virus, v, after an initial shoulder. In Model 1, the slope of the exponential decay is given by $\alpha + c$ (Fig. 7.6). Thus, the rate of CTL-mediated killing should influence the rate of free virus decline. In Model 2, however, the slope of the exponential decline is given by the smallest value among a, $\alpha + b$, and $\alpha + c$, which means that the slowest step in the life-cycle of an infected cell (including the effect of α) determines the rate of virus decline during drug treatment. If the time between infection of the cell and expressing enough viral protein to become a target for CTL is rate limiting, then the observed exponential slope of virus decline is given by a and does not depend on the rate of CTL-mediated killing. Even if a is larger than b or c, it is still possible that in most patients a is smaller than $b + \alpha$ and $c + \alpha$, and therefore, again the observed rate of virus decline reflects the initial phase of the viral life cycle and is unaffected by CTL-mediated lysis.

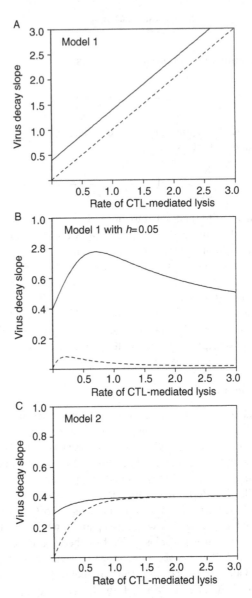

Fig. 7.6 The models provide a new interpretation of the slope of virus decay in drug treatment studies. We assume that before therapy new infections occur at a constant rate β. This leads to an equilibrium distribution of y_1, y_2 and y_3 cells and free virus v. Drug treatment reduces β to zero which leads to a decay of free virus and infected cells. (A) In Model 1, virus decline is exponential with slope $\alpha + c$ (after a shoulder of duration t_2). Thus, the observed rate of virus decline during treatment should be proportional to the rate of CTL-mediated lysis *in vivo*. (B) If we include a small fraction, h, of cells which are not exposed to CTL-mediated killing, the virus-producing cell population declines as given by eqn (7.14). In a patient with a weak CTL response

In Model 1, the equilibrium distribution of infected cells is

$$y_1(t) = 1 \quad \text{for } t < t_1,$$
$$y_2(t) = \exp[-\alpha(t - t_1)] \quad \text{for } t_1 < t < t_2, \tag{7.9}$$
$$y_3(t) = \exp[-\alpha(t_2 - t_1) - (\alpha + c)(t - t_2)] \quad \text{for } t_2 < t,$$

where t is the time since infection of the cell. Before drug treatment the total amount of virus-producing cells is

$$Y_3(0) = \int_{t_2}^{\infty} y_3(t)\,dt = [1/(\alpha + c)]\exp[-a(t_2 - t_1)]. \tag{7.10}$$

During drug treatment this cell population declines as

$$Y_3(T) = Y_3(0) \quad \text{for } T < t_2,$$
$$Y_3(T) = Y_3(0)\exp[-(\alpha + c)(T - t_2)] \quad \text{for } T > t_2. \tag{7.11}$$

Here T denotes time after start of drug treatment. Thus, virus decline occurs with a shoulder of length t_2 followed by an exponential decline with slope $\alpha + c$.

In Model 2, virus-producing cells decline as

$$Y_3(T) = \frac{b - c}{a}e^{-at} + \frac{\alpha - a + c}{\alpha + b}e^{-(\alpha+b)t} - \frac{\alpha - a + b}{\alpha + c}e^{-(\alpha+c)t}. \tag{7.12}$$

This expression again describes an initial shoulder followed by an exponential decay. The slope of the exponential decay is determined by the smallest value among a, $b + \alpha$ or $c + \alpha$. If the rate, a, at which infected cells proceed to become targets for CTL killing is slow, then the exponential decay in treatment studies may simply reflect this process and not depend on the rate of CTL-mediated killing, α.

In all models, free virus is produced from infected cells according to

$$\dot{v} = kY_3 - uv. \tag{7.13}$$

If free virus turnover is fast then $v(T)$ is proportional to $Y_3(T)$ and the decline of $Y_3(T)$ can directly be interpreted as free virus decline; if not then one more integration is

---

($\alpha \approx 0$) the exponential decline is c, in a patient with a strong CTL response ($\alpha \gg c$) the decline is again roughly c. Thus, the rate of virus decline does not reflect the rate of CTL-mediated killing, α. (C) In Model 2, virus-producing cells decline as given by eqn (7.12). This expression again describes an initial shoulder followed by an exponential decay. The slope of the exponential decay is determined by the smallest value among a, $b + \alpha$ or $c + \alpha$. If the rate, a, at which infected cells proceed to become targets for CTL killing is slow, then the exponential decay in treatment studies may simply reflect this process and not depend on the rate of CTL-mediated killing, α. For Model 1 we chose $c = 0.4$, $t_1 = 0.5$, $t_2 = 1$ and $h = 0.05$. For Model 2 we chose $a = 0.4$, $b = 2$, $c = 0.5$ (continuous lines). Broken lines indicate non-cytopathic virus, $c = 0.01$.

necessary, but the conclusions are unaffected as long as u is not the slowest rate constant which is very unlikely.

There is an additional reason why CTL-mediated killing may not affect the rate of virus decline during drug therapy. It is natural to assume that, in a given patient, infected cells are exposed to different rates of CTL-mediated lysis. This can be a consequence of spatial heterogeneity, cell tropism or antigenic variation. Therefore, rather than assuming a fixed average rate, α, of CTL-mediated lysis for all cells, it may be better to consider a distribution of different α values. In the simplest model, we assume that in each patient there is a small and variable fraction h of cells not exposed to CTL killing. Using the framework of Model 1 and assuming that a fraction, h, of cells has death rate c and the remaining $1 - h$ cells have death rate $c + \alpha$, we find that virus decay in treatment studies is given by

$$Y_3(T) = \frac{1-h}{\alpha + c} \exp[-\alpha(t_2 - t_1) - (\alpha + c)(T - t_2)] + \frac{h}{c} \exp[-c(T - t_2)]. \quad (7.14)$$

For a patient with a weak CTL response ($\alpha \approx 0$) the slope is c; for a patient with a strong CTL response ($\alpha \gg c$) the slope is again dominated by c. Therefore, if a small fraction of infected cells (less than 5–10%) escape from CTL surveillance, the resulting virus decay slopes are independent of the average rate of CTL-mediated killing in this patient. The intuitive reason behind this explanation is that virus dynamics experiments measure the turnover rate of cells that produce most of the plasma virus. But cells which are killed by CTL produce less plasma virus than cells that are not killed by CTL. Therefore, the virus decay slope will be biased toward the decay rate of cells which have escaped from CTL-mediated lysis. Clearly, the same argument applies to Model 2.

7.4   Comparing HIV and HBV

There is some discussion as to whether HIV is cytopathic *in vivo*. Figure 7.7 shows the slope of virus decline during drug therapy as a function of the rate of CTL-mediated lysis for both a cytopathic virus (continuous line) and a non-cytopathic virus (broken line). For Model 1, the decay slope is a linear function of the rate of CTL-mediated lysis. Thus patients with a stronger CTL response should have a faster half-life of productively infected cells. This is not observed in HIV-1 infection; the half-life is roughly the same for all patients (Fig. 7.1).

For both Model 1 with heterogeneity and Model 2, we find that the virus decay slope is roughly constant (and reasonably fast) for a cytopathic virus. For a non-cytopathic virus, the observed decay of productively infected cells during treatment should be very slow for Model 1 with heterogeneity and very dependent on the rate of CTL-mediated lysis in case of Model 2.

Therefore, the lack of extensive variation in virus decay slopes and the short half-life of productively infected cells in HIV infected patients undergoing drug therapy, strongly suggest that HIV-1 is cytopathic *in vivo*. Otherwise one would have to argue that all individuals measured so far have a rather constant CTL (or immune)-mediated clearance rate of infected cells that corresponds to a target cell lysis of 5–10% in 4 h, which seems unlikely. If HIV-1 were non-cytopathic there should be much smaller decay rates in patients with weak or absent CTL responses and a greater heterogeneity of decay slopes.

The argument gains further support by comparison with the results obtained for hepatitis B virus (Chapter 5). For HBV the half-life of productively infected cells varies from 10 to 100 days in different individuals. Thus, there is considerable heterogeneity in the half-lives and they are longer than in HIV-1 infection. This is consistent with the notion that HBV is largely non-cytopathic *in vivo*.

With respect to HIV-1 disease progression, the crucial question is to what extent immune responses reduce virus load in infected patients. Our results suggest that CTL-mediated lysis is sufficiently fast to eliminate a large fraction of productively infected cells and thereby greatly reduce virus production. In addition, CD8+ T cells also release cytokines that can block virus entry into cells. Both mechanisms are likely to reduce virus load and, therefore, slow down the rate of disease progression.

7.5   Further reading

McMichael and Phillips (1997) provide a review of CTL activity in HIV infection. Many groups have characterized anti-HIV CTL responses; the data summarized in Fig. 7.4 are from Walker *et al.* (1987, 1988), Plata *et al.* (1987), Koenig *et al.* (1988), Gotch *et al.* (1990), McElrath *et al.* (1994) and Buseyne *et al.* (1993). The data shown in Fig. 7.5 are from Rinaldo *et al.* (1995a,b). Klenerman *et al.* (1996a) provide a mathematical model for the effect of CTL-mediated lysis on virus dynamics. For immune responses during primary HIV infection see Borrow *et al.* (1994), Koup *et al.* (1994).

8

WHAT IS A QUASISPECIES?

The word quasispecies is a mystery for most biologists, possibly because it was invented by chemists. In the 1970s, Manfred Eigen and Peter Schuster developed a chemical theory for the origin of life. They described how populations of RNA molecules could reproduce themselves. They noted that the spontaneous chemical reproduction of such comparatively simple molecules was much less accurate than the genetic replication of any organism currently alive. The first proliferation of biological information was thus an extremely error-prone process, compared to the more accurate replication mechanisms which appear in modern cells. Random events would lead to mutations. The reproduction of such molecules by base-pairing is not completely accurate, so a certain degree of mismatching is to be expected. Consequently a population of RNA molecules that was the result of such an inaccurate replication process would not be absolutely homogeneous, but a mixture of RNA molecules with different nucleotide sequences.

Chemists refer to an ensemble of equal molecules as 'species'. For example, the 'species' of all H_2O molecules. In contrast, a species of RNA molecules, derived by inaccurate reproduction, is not an ensemble of *identical* molecules. Hence the term 'quasispecies'. For biologists, the term is confusing because a biological species is a complicated and in some sense loosely defined concept. How much more unspecific is then a 'quasispecies'!

Eigen and Schuster were primarily interested in the origin of life. They assumed that RNA was the first biological replicator. In the primordial soup, the four nucleotides (adenine, guanosine, cytidine and thymidine) would spontaneously form short chains, so called polymers. These polymers could in principle reproduce by base-pairing. Alongside of each polymer another polymer would form which consisted of the complimentary sequence of nucleotides. Subsequently, the double helix could split up and the two single strands would go on to form new double helices, thereby imprinting their sequences on new polymers. This process, which is now the basis of all life and is conducted by highly sophisticated and accurate enzymes inside the cell, might have been occuring initially very slowly and subject to high error-rates.

This primitive genetic replication is a chemical process and can be described by chemical kinetics, that is by equations specifying how the concentration of certain molecules change over time. The kinetics of RNA self-replication assume that molecules have different replication rates according to their sequence. This means that some may produce 'offspring' faster than others; they are 'fitter'. In addition the theory takes into account that replication is inaccurate. An offspring sequence need not be identical to its parent, but may differ in certain positions. A substitution of one base for another is called a 'point mutation.' The equations do not lead to a population of identical sequences but to an ensemble of related but different sequences. This ensemble is called quasispecies.

More precisely, Eigen and Schuster refer to the *equilibrium distribution* of sequences that is formed by this mutation and selection process as quasispecies. 'Mutation' because reproduction is subject to errors. 'Selection' because sequences have different fitnesses. Eigen and Schuster go on to argue that the target of natural selection is not the fittest sequence, but the quasispecies. Natural selection will not just chose the fittest type, but an ensemble of different variants. The fittest sequence may only represent a very small fraction of the quasispecies; it may indeed not be present at all.

An important concept of quasispecies theory is the *error-threshold*. If the mutation rate is too high, that is if too many mistakes occur in any one replication event, then the population will be unable to maintain any genetic information. In the long run the composition of the quasispecies will only be determined by randomness. The abundance of individual sequences will be independent of their fitness. Thus the error-rate must be below a critical threshold level for the system to maintain information. If the error-rate is expressed as a per base probability to make a mistake, then the error-threshold can be written as a condition which limits the maximum length of the RNA sequence. For any given per base fidelity the sequence can only reach a certain length. Eigen and Schuster estimated that this critical length should be roughly 100 bases for self-replicating RNA molecules in the primordial soup.

Since every biological reproduction is error-prone, the quasispecies concept can be readily applied to genetic processes other than RNA self-replication. Populations of viruses, bacteria, plants or animals are quasispecies. Their genetic reproduction is of course more complicated than a simple copying of the sequence and will include more sophisticated mutational events (such as recombination, or sexual reproduction), but the underlying principles remain the same. Any natural biological population will be a mix of genomes, a quasispecies.

The error-threshold concept can also be expanded to more complicated organisms. Very roughly, it can be shown that the genome length should not exceed one over the per base mutation rate. Viral RNA replication, in the absence of any error-correcting mechanisms (proof-reading), has a per base mutation rate of about 10^{-4}. As predicted, the genome length of such viruses is thus about 10^4. The human genome is about 3×10^9 bases long, and the high quality DNA replication enzymes in human cells ensure an error rate of about 10^{-9} per base. The prediction that the genome length does not exceed one over the per base mutation rate holds for many different organisms.

8.1 More than atoms in our universe

Consider an RNA or DNA sequence of length l denoting the number of bases in the sequence. There are 4^l different variants. This means that even for moderate lengths a 'hyper-astronomically' large number of different variants can be formed. For example, for a polynucleotide of length $l = 300$, which is just large enough to encode for one of the smallest proteins, there are more than 10^{180} different variants. The genome length of HIV is about 10 000 bases. A particular HIV sequence is one choice out of 10^{6020}

different nucleotide sequences of the same length. For comparison, there are only 10^{11} stars in our galaxy, or about 10^{80} protons in our universe.

Suppose that the average virus load in an (untreated) HIV infected person is 10^{10} particles and that this amount of virus particles is produced every day. Thus in 10 years, an HIV infection could generate about 10^{13} virus particles. All HIV infected people in the world could generate up to 10^{20} virus particles over the course of 10 years. If we assume that only 1% of nucleotides in the HIV genome are variable (which is a gross underestimation), then there would be about 10^{60} HIV mutants that coincide in 99% of their genetic sequences. Thus the worldwide pandemic would only produce a fraction 10^{-40} of all such variants during 10 years. While HIV is extremely variable, at any one time only a minute fraction of all possible variants is present worldwide.

8.2   Quasispecies live in sequence space

The correct geometry for quasispecies is not the usual Euclidian space but the so-called sequence space. In the sequence space, all possible variants of a given length are arranged such that neighbours differ by only one base substitution. More generally, the distance between two sequences equals the number of substitutions between them. For example, the sequence AATCG differs from ATCCG in two positions. The dimension of the space is given by the length of the sequence. In each dimension there are four possibilities, corresponding to the four nucleotides, A, T, C and G. The sequence space that contains all sequences of length five has five dimensions and $4^5 = 1024$ points (different sequences).

The important features of this sequence space are: (1) its high dimensionality, (2) the large number of shortest mutational routes between two distant mutant sequences (for two sequences separated by d point mutations there are $d!$ shortest mutational routes) and (3) that many sequences are confined to a close neighbourhood of each other. The diameter of a sequence space that contains 10^{80} points is only 133 'length units', i.e. point mutations. This means that relatively few point mutations can lead from one region in the sequence space to a completely different region providing there exists something like a guiding gradient to avoid going in 'wrong directions'. In evolution, this gradient is provided by natural selection.

8.3   Quasispecies explore fitness landscapes

Every point in sequence space can also be assigned a fitness value (representing the reproduction rate of this sequence). This leads to the concept of a fitness landscape. Fitness landscapes have one more dimension than the corresponding sequence space, because for every point there is a 'height'. Quasispecies wander over the fitness landscapes searching for peaks, which represents region of high fitness values. Under the guidance of natural selection quasispecies climb the mountains in the high dimensional fitness landscape.

Here again we can easily envision how natural selection does not simply choose the fittest sequence, but the quasispecies. Imagine two sequences A and B. Assume A has a higher replication rate than B, thus it has a higher intrinsic fitness value. Suppose A

is surrounded by mutants with very low fitness, while B is surrounded by mutants with high fitness. (Both A and B are local optima, but A is a very sharp peak in the fitness landscape, whereas B is the top of a flat mountain.) In the absence of mutation, A will be selected and B will disappear. With mutation, however, the situation can change and B could be the winner. In fact, the mathematical equations will show a critical mutation rate, below which A is the winner, but above which B and its neighbours are favoured.

The natural mathematical habitat for quasispecies is the sequence space. A quasispecies is defined by assigning abundances to every possible sequence (to every point of the sequence space). For reasonable genome lengths, most of the mutants will have zero abundance, because the number of possible sequences usually exceeds the population size. A quasispecies is a small cloud in sequence space moving under the influence of mutation and selection.

8.4 The mathematics of quasispecies

Suppose there are n different nucleic acid sequences $v_1, v_2, \ldots, v_n$. Each variant is characterized by a specific nucleotide sequence, which determines the replication rate of a given variant. Denote the replication rates of the variants $v_1, v_2, \ldots, v_n$ by $a_1, a_2, \ldots, a_n$. These quantities represent the selective values of the individual mutants. In the absence of mutation the variant with the highest replication rate will grow fastest and reach fixation. The result of selection in this world without errors is a homogeneous population consisting of the fastest replicating variant. But replication is not error-free. Thus it is necessary to define the probabilities Q_{ij} that (erroneous) replication of template v_j results in the production of the sequence v_i. The quantities Q_{ij} for $i = 1, 2, \ldots, n$ and $j = 1, 2, \ldots, n$ form the so called mutation matrix.

Quasispecies replication can be visualized by the following chemical reactions:

$$\text{error-free:} \quad (A) + v_i \xrightarrow{a_i Q_{ii}} 2v_i,$$

$$\text{mutation:} \quad (A) + v_i \xrightarrow{a_i Q_{ji}} v_i + v_j.$$

The symbol A denotes low-molecular-weight materials (the four nucleotides) which are required for DNA or RNA synthesis. It is assumed that the available amount of A is constant and hence it will not enter as a variable into the kinetic differential equations. Error-free replication and mutation are parallel reactions of the same mechanism. The rate of replication, a_i, depends on the template I_i; the mutation probability, Q_{ij}, on both the template and the product of replication.

A system of ordinary differential equations describes the time evolution of the population of these nucleic acid sequences. The growth rate of a specific variant, e.g. v_1, can be written as:

$$\dot{v}_1 = a_1 Q_{11} v_1 + a_2 Q_{12} v_2 + \cdots + a_n Q_{1n} v_n. \tag{8.1}$$

New particles of variant v_1 can be formed by error-free replication of v_1; this happens at the replication rate a_1 and the probability Q_{11}; the overall rate is therefore given by $a_1 Q_{11} v_1$. Erroneous replication of any other mutants $v_2, \ldots, v_n$ can also lead to new v_1

particles. This is represented by the growth terms $a_2 Q_{12} v_2 + \cdots + a_n Q_{1n} v_n$ in eqn (8.1). In the same way we can write the rate of production of any of the other variants to obtain the whole system of differential equations:

$$\dot{v}_i = \sum_{j=1}^{n} a_j Q_{ij} v_j, \quad i = 1, \ldots, n. \tag{8.2}$$

In this context the population will no longer consist only of the fastest growing sequence, but a whole ensemble of mutants with different replication rates. This ensemble of mutants is the quasispecies.

The frequency of a given variant within the quasispecies does not depend on its replicative value alone, but also on the likelihood with which it is produced by erroneous replication of other templates and their frequencies in the quasispecies distribution. This is important to the understanding of the structural organization of a quasispecies. The consequence of this effect is that the individual sequence v_i with its replicative value a_i no longer serves as the unit (or target) of selection. The quasispecies itself is the target of selection in a mutation–selection process. This fact has important implications. Evolution is normally thought of as the interaction between mutation and selection. Selection is a factor that favours advantageous mutants that have been generated by pure chance; indeed, it is normally considered a mistake to think of mutations as being guided other than by chance. A quasispecies, however, can guide mutations. This does not mean that there is any correlation between the (intrinsically stochastic) act of mutation and the selective advantage of the mutant. But selection operates on the structure of the whole quasispecies which is adapted to its *fitness lanscape* (this term is originally from Sewall Wright). Therefore evolution can be guided towards the peaks of this fitness landscape. This happens because more successful mutants (that may be in closer neighborhood to the peaks of the landscape) will produce more offspring than less successful mutants (which may be further away of the peaks). Evolutionary optimization can be viewed as a hill-climbing process of the quasispecies that occurs along certain pathways in sequence space.

8.5 Error-thresholds

Another important concept in quasispecies theory is the error-threshold of replication. If replication were error-free, no mutants would arise and evolution would stop. Evolution would, however, also be impossible if the error-rate of replication were too high (only some mutations may lead to an improvement in adaptation, but most will lead to deterioration). The quasispecies concept allows us to quantify the resulting minimal replication accuracy that maintains adaptation.

Let us assume a population consists of (1) a fast replicating variant v_1—the wild-type sequence—with replication rate a_1 and (2) its mutant distribution (error-tail) v_2 with a lower average replication rate a_2. Let q denote the per base accuracy of replication, i.e. the probability that a single base is accurately replicated. Thus the probability that the whole sequence (of length l) is replicated without errors is given by $Q = q^l$. Neglecting the small probability that erroneous replication of a mutant gives rise to a wild-type sequence leads to the equations:

$$\dot{v}_1 = a_1 Q v_1,$$
$$\dot{v}_2 = a_1 (1 - Q) v_1 + a_2 v_2. \tag{8.3}$$

Here the ratio of wild-type over mutants converges to

$$\frac{v_1}{v_2} \rightarrow \frac{a_1 Q - a_2}{a_1 (1 - Q)}. \tag{8.4}$$

Therefore the wild-type can only be maintained in the population if $Q > a_2/a_1$. This means that the single digit replication accuracy, q, must be larger than a certain critical value. This error-threshold relation is obtained as

$$q > q_{\text{crit}} = \left(\frac{a_2}{a_1} \right)^{1/l}. \tag{8.5}$$

For replication accuracies lower than q_{crit} the wild-type sequence will be lost from the population although it has the highest replication rate. This leads to an important relationship between the replication accuracy and the sequence length

$$l < \frac{1}{1 - q}. \tag{8.6}$$

Here we have used the approximation that the logarithm of a_1/a_2 is about 1. The inequality (8.6) represents an approximation for the upper genome length l that can be maintained by a given single digit replication accuracy without loosing adaptation.

8.6 Some fancier quasispecies maths

The standard equation of quasispecies theory is

$$\frac{dv}{dt} = Wv - d(v)v. \tag{8.7}$$

The vector, v, contains the population sizes of the individual sequences $v = (v_1, v_2, \ldots, v_n)$. The matrix W contains the replication rates and mutation probabilities

$$W = \begin{pmatrix} a_1 Q_{11} & a_2 Q_{12} & \cdots & a_n Q_{1n} \\ a_1 Q_{21} & a_2 Q_{22} & \cdots & a_n Q_{2n} \\ \vdots & \vdots & \ddots & \vdots \\ a_1 Q_{n1} & a_2 Q_{n2} & \cdots & a_n Q_{nn} \end{pmatrix}. \tag{8.8}$$

The mutation matrix is given by:

$$Q_{ij} = p^{H_{ij}} (1 - p)^{(l - H_{ij})}. \tag{8.9}$$

Here p is the mutation rate per bit, l is the length of the bitstring, and H_{ij} is the Hamming distance between strains i and j, that is the number of bits in which the two strains differ. Error-free replication is given by $Q_{ii} = (1 - p)^l$.

In addition one usually includes an unspecific degradation term, $v_i \to 0$, which may be adjusted in such a way that the total population is of constant size. This is achieved by choosing

$$d(x) = \sum_{i=1}^{n} a_i x_i \Big/ \sum_{i=1}^{n} x_i. \tag{8.10}$$

The equilibrium of eqn (8.7) can be calculated by solving the standard eigenvalue problem of linear algebra

$$Wx = \lambda x. \tag{8.11}$$

The quasispecies is then given—in precise mathematical terms—as the dominant eigenvector v^* which belongs to the largest eigenvalue λ_{max} of the matrix W. This eigenvector, v^*, describes the exact population structure of the quasispecies; each mutant, i, is contained in the quasispecies with frequency v_i^*. (We can normalize such that $\sum v_i^* = 1$.) The largest eigenvalue is exactly the average replication rate of the quasispecies, $\lambda_{max} = \sum a_i v_i^*$.

8.7   Viral quasispecies

Let us now embed the quasispecies into the basic model of virus dynamics. Denote by x, y_i and v_i, respectively, uninfected cells, cells infected by virus mutant i and free virus particles of mutant i. We have

$$\begin{aligned}
\dot{x} &= \lambda - dx - x \sum_i \beta_i v_i, \\
\dot{y}_i &= x \sum_j Q_{ij} \beta_j v_j - a_i y_i, \\
\dot{v}_i &= k_i y_i - u_i v_i.
\end{aligned} \tag{8.12}$$

As in Chapter 3, uninfected cells are produced at a rate, λ, and die at a rate d. Infected cells of type i are produced when an uninfected cell is infected by a virus particle of type i. This happens at rate β_i. Infected cells produce free virus particles at rate k_i. Free virus particles and infected cells die at rates u_i and a_i, respectively. Q_{ij} is the probability of strain j mutating to strain i. Without mutation, the basic reproductive ratio of viral strain i, is given by $R_i = \lambda b_i k_i / (\delta a_i u_i)$. Strains are only viable if $R_i > 1$, and the strain with the largest R_i will outcompete all other strains.

With mutation, there is a critical error-rate, p_c, beyond which the strain with the highest R_i fails to be selected. Let us consider a single peak fitness landscape, where strain 1 has the highest basic reproductive ratio, R_1, and all other strains have the same, but lower basic reproductive ratio R. Neglecting back mutation the critical error-rate is given by $p_c = (R/R_1)^{1/l}$. If $p < p_c$ the quasispecies will be centered around the fittest strain 1, which will be most abundant. If $p > p_c$, each virus strain has essentially the same relative abundance.

In the above virus model there are in fact two different types of error-thresholds. If $R < 1$, the virus population will become extinct for $p > p_c$. If $R > 1$ the virus population will survive for $p > p_c$, but the fittest strain will disappear. Therefore increasing the mutation rate of a virus with an appropriate drug can eliminate the fittest virus mutants and thereby reduce virus load; further increase of the mutation rate can drive the whole virus population to extinction.

8.8 Antigenic escape and optimum mutation rate

An interesting feature of HIV infection is the enormous genetic variability of the virus found in any one patient at any one time. Since the virus is usually opposed by strong immune responses, part of this genetic variation is to escape from these immune responses. Chapters 12–15 will deal with detailed models of virus evolution and antigenic variation. Here we simply ask the question what is the optimum mutation rate of the virus to maximize the probability of antigenic escape.

Denote by l the total length of the virus genome; for HIV we have $l \approx 10^4$. Let m denote the number of sites within the viral genome such that mutation in at least one of these sites results in the production of an escape mutant. The probability to obtain an escape mutant is give by $q^{l-m}(1 - q^m)$. This is simply the probability for obtaining a mutant with no errors in $l - m$ sites and at least one error in m sites. This probability has a maximum for $q = (1 - m/l)^{1/m}$ which for $m \ll l$ is well approximated by $q = 1 - 1/l$. Therefore the optimum mutation rate is given by $p = 1/l$ which for HIV is about 10^{-4}. This is in good agreement with experimental observation: Preston et al. (1988) and Roberts et al. (1988) measure the HIV-1 mutation rate to be about 10^{-4}, while Mansky and Temin (1995) obtain 3×10^{-5}.

It is interesting to note that the optimum mutation (which maximizes the probability of antigenic escape) is very close to the maximum mutation rate; the error-threshold argument requires the mutation rate not to exceed $1/l$. This principle is related to concepts like 'evolution on the edge of chaos' and 'self-organized criticality'.

8.9 Further reading

Seminal papers on quasispecies theory are Eigen and Schuster (1977), Eigen (1971), Fontana and Schuster (1987), Swetina and Schuster (1982), McCaskill (1984), Eigen et al. (1989), McCaskill (1984a), Fontana et al. (1989), and Fontana and Schuster (1998). The error-threshold in finite populations is calculated in Nowak and Schuster (1989). Maynard Smith (1970) introduces the idea of a sequence space. Experimental papers on quasispecies evolution are Mills et al. (1973), Semper and Luce (1975), Holland et al. (1982), Biebricher et al. (1985, 1986), and Duarte et al (1994). Viral quasispecies are described by Domingo et al. (1978), Domingo et al. (1980), Spindler et al. (1982), Parvin et al. (1986), and Wain-Hobson (1989). HIV diversity is studied by Hahn et al. (1986), Saag et al. (1988), Meyerhans et al. (1989), and Simmonds et al. (1990). The optimal mutation rate for antigenic variation is from Nowak (1990).

9

THE FREQUENCY OF RESISTANT MUTANT VIRUS
BEFORE ANTI-VIRAL THERAPY

A number of potent anti-HIV drugs are now available which inhibit the replication of the virus *in vivo*. Reverse transcriptase inhibitors prevent the infection of new cells, while protease inhibitors prevent infected cells from producing infectious virus particles. As discussed in Chapter 4, drug treatment usually results in a rapid decline of plasma virus load and an increase in the CD4 cell count. Monotherapy often leads to rapid emergence of drug-resistant virus mutants. For some drugs, a single-point mutation can confer high level resistance, while for other drugs several point mutations are required. Combination therapy can result in a longer lasting suppression of virus load.

The success of anti-viral therapy and the emergence of resistance depend crucially on whether or not resistant mutant virus is present before the initiation of therapy. If resistant virus is present prior to treatment then application of the drug leads to declining levels of sensitive wild-type virus and increasing levels of resistant mutant virus. Mutant virus grows either because of an increased supply of target cells or a reduction in anti-viral immunity.

We call a virus mutant *resistant* if it has a positive growth rate in the presence of therapy. More precisely, we ask that its basic reproductive ratio, R_0, during treatment is greater than unity. As defined in Chapter 3, the basic reproductive ratio is the number of productively infected cells that arise out of any one productively infected cell when the supply of target cells is not reduced. Thus a mutant is *resistant* if its R_0 during treatment exceeds one. Conversely, a mutant is *sensitive* if its R_0 during treatment is less than one.

This definition of drug resistance has important consequences: resistance is an *in vivo* feature of the virus; it cannot solely be determined by *in vitro* measurements of drug sensitivity. Resistance depends on the supply and permissiveness of target cells and on the strength of the patient's immune response against the virus. Resistance depends on both the virus and the host. The same mutant could be resistant in one patient, but sensitive in another patient.

For monotherapy, it is likely that 1- or 2-point mutations confer resistance. Hence mutants resistant to monotherapy are expected to be present in many HIV infected people. Resistance to two or three drugs, however, requires several point mutations, and such mutants may not be present in untreated patients. Whether or not a particular mutant virus is present in a patient depends on the selective disadvantage of this mutant (and its neighbours in sequence space) and the mutation rate of virus replication.

In this chapter, we will use a quasispecies equation of viral dynamics to calculate the expected frequency of drug-resistant virus in untreated patients. We will provide results for mutants that differ in 1-, 2-, 3- or more-point mutations from wild-type virus. We will compare the effect of mutation and selection on the pre-treatment frequency of resistant virus.

For example, we will show that the equilibrium ratio of a mutant virus that differs from wild-type virus in 2-point mutations is given by

$$\frac{y_{11}^*}{y_{00}^*} = \left(\frac{\mu_1\mu_2}{s_{11}}\right)\left(\frac{1}{s_{01}} + \frac{1}{s_{10}} - 1\right).$$

Here y_{11}^* and y_{00}^* denote the equilibrium abundance of the 2-error mutant and the wild-type virus, respectively. The selective disadvantage of the 2-error mutant is given by s_{11}, while s_{01} and s_{10} define the selective disadvantages of the two intermediate 1-error mutants. The mutation rate of viral replication in the two relevant positions is given by μ_1 and μ_2.

More generally, for the equilibrium ratio, ρ, of an n-error mutant prior to treatment we find:

$$\rho = n!\left(\frac{\mu}{s}\right)^n.$$

Here we have assumed that all mutants have the same selective disadvantage, s, and that the mutation rate of viral replication is the same for all n positions and given by μ. If, for example, we set $\mu = 3 \times 10^{-5}$, which is thought to be the average mutation rate of the HIV reverse transcriptase enzyme, and $s = 0.1$, which implies that the selective disadvantage of the resistant n-error mutant and all intermediate mutants is 10%, we find for the equilibrium frequency of a 1-error mutant 3×10^{-4}, of a 2-error mutant 2×10^{-7}, and of a 3-error mutant 2×10^{-10}. Suppose there are about 10^8 productively infected cells in a patient. If resistance is conferred by a particular 1- or 2-error mutant then, in this numerical example, we would expect the patient to have resistant virus prior to treatment. If 3-point mutations are required it becomes a matter of chance, whereas if 4- or more-point mutations are necessary then the patient has most likely no resistant virus prior to treatment.

In Section 9.1, we discuss the mathematical model for the simple case where wild-type and mutant virus differ in only 1-point mutation. In Section 9.2, we turn to 2-point mutations and in 9.3 to n-point mutations. Section 9.4 provides some practical implications.

9.1   Wild-type and mutant differ by 1-point mutation

We begin with a simple model that contains wild-type virus and one mutant. We assume that wild-type and mutant differ only by a single-point mutation. The mutation rate is given by the parameter μ. We further assume that wild-type virus has a fitness advantage over mutant virus in the absence of drug; the selection coefficient is given by s. Thus if the fitness of the wild-type is 1, the fitness of the mutant is $1 - s$.

The model has three variables: uninfected cells, x, cells infected by wild-type virus, y_0, and by mutant virus, y_1. Uninfected cells are produced at rate λ and die at rate

d. Infected cells die at rate a. Infected cells give rise to new infected cells at a rate proportional to the abundance of uninfected cells times infected cells; the rate constant is β. We assume that free virus dynamics is fast compared to infected cell turnover. Therefore we do not consider a separate equation for free virions, but simply assume that virion abundance is proportional to infected cell abundance. The model equations are:

$$\dot{x} = \lambda - dx - \beta x[y_0 + (1-s)y_1],$$
$$\dot{y}_0 = \beta x[(1-\mu)y_0 + (1-s)\mu y_1] - ay_0,$$
$$\dot{y}_1 = \beta x[\mu y_0 + (1-s)(1-\mu)y_1] - ay_1.$$

At equilibrium the ratio of mutant to wild-type virus, to first order in μ, is

$$\frac{y_1^*}{y_0^*} = \frac{\mu}{s}.$$

For example, if the point mutation rate is $\mu = 3 \times 10^{-5}$ and the mutant has a selective disadvantage of $s = 0.01$, then the relative proportion of mutant to wild-type is 3×10^{-3}. In other words, for this choice of parameters, about 1 in 300 cells contains the mutant virus. There is some experimental evidence for certain 1-error mutants to occur with frequencies of the order of 10^{-3} in untreated patients (Nájera *et al.* 1995). If, on the other hand, the mutant has a large selective disadvantage (that is $s \approx 1$) then the ratio of mutant to wild-type is given by the mutation rate, μ. In this case we would expect about 1 in 30 000 infected cells to contain the mutant.

9.2   Wild-type and mutant differ by 2-point mutations

We now expand the model to mutants which differ by 2-point mutations. This model has 5 variables: uninfected cells, x, cells infected by wild-type virus, y_{00}, cells infected by 1-error mutants, y_{01} and y_{10}, and cells infected by the 2-error mutant, y_{11}. Note that we use a binary notation for wild-type virus (00) and the various mutants (01, 10, and 11). Thus at each of two relevant positions we consider one amino acid substitution. We assume that these amino acid substitutions are due to a single base substitution in the viral genome. The replication fidelity of the reverse transcriptase may vary for different positions in the genome and therefore we assume that the mutation rate for the first position is μ_1 and for the second position μ_2. The selective disadvantage of the mutants y_{01}, y_{10} and y_{11} is given by s_{01}, s_{10} and s_{11}, respectively. The model equations are:

$$\dot{x} = \lambda - dx - \beta x(y_{00} + r_{01}y_{01} + r_{10}y_{10} + r_{11}y_{11}),$$
$$\dot{y}_{00} = \beta x(v_1 v_2 y_{00} + r_{01}v_1\mu_2 y_{01} + r_{10}\mu_1 v_2 y_{10} + r_{11}\mu_1\mu_2 y_{11}) - ay_{00},$$
$$\dot{y}_{01} = \beta x(v_1\mu_2 y_{00} + r_{01}v_1 v_2 y_{01} + r_{10}\mu_1\mu_2 y_{10} + r_{11}\mu_1 v_2 y_{11}) - ay_{01},$$
$$\dot{y}_{10} = \beta x(\mu_1 v_2 y_{00} + r_{01}\mu_1\mu_2 y_{01} + r_{10}v_1 v_2 y_{10} + r_{11}v_1\mu_2 y_{11}) - ay_{10},$$
$$\dot{y}_{11} = \beta x(\mu_1\mu_2 y_{00} + r_{01}\mu_1 v_2 y_{01} + r_{10}v_1\mu_2 y_{10} + r_{11}v_1 v_2 y_{11}) - ay_{11}.$$

We have used the abbreviations $v_i = 1 - \mu_i$ and $r_{ij} = 1 - s_{ij}$. The exact equilibrium solution of this system is complicated, but we can obtain a good approximation by neglecting back-mutations. In this case, we simply solve the eigenvalue equation:

$$
\begin{bmatrix}
\nu_1\nu_2 & 0 & 0 & 0 \\
\nu_1\mu_2 & r_{01}\nu_1 & 0 & 0 \\
\nu_2\mu_1 & 0 & r_{10}\nu_2 & 0 \\
\mu_1\mu_2 & r_{01}\mu_1 & r_{10}\mu_2 & r_{11}
\end{bmatrix}
\times
\begin{bmatrix}
y_{00} \\ y_{01} \\ y_{10} \\ y_{11}
\end{bmatrix}
= \Lambda \times
\begin{bmatrix}
y_{00} \\ y_{01} \\ y_{10} \\ y_{11}
\end{bmatrix}.
$$

We obtain

$$
\frac{y_{01}^*}{y_{00}^*} = \frac{\mu_2}{s_{01} - \mu_2},
$$

$$
\frac{y_{10}^*}{y_{00}^*} = \frac{\mu_1}{s_{10} - \mu_1},
$$

$$
\frac{y_{11}^*}{y_{00}^*} = \frac{\mu_1\mu_2(\mu_1\mu_2 - \mu_1 - \mu_2 + s_{01} + s_{10} - s_{01}s_{10})}{(s_{01} - \mu_2)(s_{10} - \mu_1)(\mu_1\mu_2 - \mu_1 - \mu_2 + s_{11})}.
$$

If in addition we assume that the selection coefficients are larger than the mutation rates, then we obtain the approximate equilibrium ratios:

$$
\frac{y_{01}^*}{y_{00}^*} = \frac{\mu_2}{s_{01}},
$$

$$
\frac{y_{10}^*}{y_{00}^*} = \frac{\mu_1}{s_{10}},
$$

$$
\frac{y_{11}^*}{y_{00}^*} = \left(\frac{\mu_1\mu_2}{s_{11}}\right)\left(\frac{1}{s_{01}} + \frac{1}{s_{10}} - 1\right).
$$

For example, if $\mu_1 = \mu_2 = 3 \times 10^{-5}$ and if $s_{01} = s_{10} = s_{11} = 0.01$ then $y_{11}/y_{00} \approx 2 \times 10^{-5}$. In other words, about 1 in 50 000 productively infected cells contains resistant mutant virus, which differs from the wild-type by 2-point mutations.

In Table 9.1, we show the effect of different selective disadvantages for the intermediate mutants, s_{01} and s_{10}, on the relative frequency of y_{11}. For example, if $s_{11} = 0.01$, but $s_{01} = s_{10} = 1$ (i.e., the double mutant has a small selective disadvantage, but the intermediate 1-error mutants cannot replicate) then the relative abundance of the double mutant is $y_{11}/y_{00} \approx 9 \times 10^{-8}$. If, on the other hand, $s_{01} = s_{10} = 0.001$ (i.e., the intermediate mutants have a selective disadvantage of only 0.1% compared to wild-type) then we find $y_{11}/y_{00} \approx 1.8 \times 10^{-4}$. The difference in the equilibrium frequency of the double mutant is several orders of magnitude. Therefore, in order to estimate the pre-treatment equilibrium frequency of a particular resistant mutant it is not sufficient to know the selective disadvantage of this mutant compared to the wild-type, but it is also necessary to know the selection coefficients of the intermediate mutants.

9.3 Wild-type and mutant differ by n-point mutations

The above model can be expanded to include mutants that differ from wild-type virus by more than two point mutations. For simplicity, assume that the average mutation rate

Table 9.1 *Effect of the selective disadvantages of the intermediate 1-error mutants,* y_{01} *and* y_{10}, *on the relative abundance of the 2-error mutant,* y_{11}

s_{01}, s_{10}	10^{-3}	10^{-2}	10^{-1}	1
10^{-3}	1.8×10^{-4}	9.9×10^{-5}	9.1×10^{-5}	9.0×10^{-5}
10^{-2}		1.8×10^{-5}	9.9×10^{-6}	9.0×10^{-6}
10^{-1}			1.7×10^{-6}	9.0×10^{-7}
1				9.0×10^{-8}

We assume a mutation rate of $\mu = 3 \times 10^{-5}$ and a selective disadvantage for y_{11} of $s_{11} = 0.01$. We vary the selective disadvantages, s_{01} and s_{10}, of the intermediate mutants between 0.001 and 1. The table shows the equilibrium frequency, relative to wild-type, of the resistant y_{11} mutant. Since the effect of s_{01} and s_{10} is the same, we only present half of the symmetric table.

for each site is the same, μ, and that all mutants have the same selective disadvantage, s, which is much smaller than one. In this case we obtain for an n-error mutant

$$\frac{y_{11...1}^*}{y_{00...0}^*} = n! \left(\frac{\mu}{s}\right)^n.$$

Of course, $n!(\mu/s)^n$ has to be less than unity otherwise our approximations break down. Observe that this is essentially an error-threshold condition of quasispecies theory.

Table 9.2 shows equilibrium frequencies of mutants that differ from wild-type by up to 5-point mutations. For $s = 0.01$, the frequency of a 3-error mutant is 1.6×10^{-7}, of a 4-error mutant is 1.9×10^{-9} and of a 5-error mutant 2.9×10^{-11}. Suppose that there are about 10^8 productively infected cells in an HIV-1 infected patient. For $s = 0.01$, a particular 4- or 5-error mutant is unlikely to be present, while the 3-error mutant is likely to exist. For a higher selective disadvantage ($s > 0.01$) also the 3-error mutant is unlikely to exist in a patient prior to treatment. We can also ask the question, which maximum selective disadvantage is still compatible with survival of a given virus mutant in a patient with N infected cells. If $N = 10^8$ and $\mu = 3 \times 10^{-5}$, then a 2-error mutant will survive if $s < 0.4$, a 3-error mutant if $s < 0.03$ and a 4-error mutant if $s < 0.007$. In this calculation, intermediate mutants have the same selective disadvantage s. Other assumptions are also possible and can be calculated.

The model also provides insight into the time evolution of resistant mutant virus during the course of infection. If a patient is initially infected by sensitive wild-type virus, y_0, then the frequency of resistant mutant virus, y_1, rises as:

$$\frac{y_1(t)}{y_0(t)} = \left(1 - \exp^{-\beta x(s-\mu)t}\right) \frac{\mu}{(s - \mu)}.$$

Note that this calculation assumes that the number of uninfected cells, x, is approximately constant during the rise of resistant mutants. The mutant will arrive after about $\ln 2/[\beta x(s - \mu)]$ days at half its equilibrium value. In principle, we can recursively

Table 9.2 *Equilibrium frequencies of mutants that differ from the wild-type by 1-, 2-, 3-, 4- or 5-point mutations*

n	$s = 0.001$	$s = 0.01$	$s = 0.1$
1	3×10^{-2}	3×10^{-3}	3×10^{-4}
2	1.8×10^{-3}	1.8×10^{-5}	1.8×10^{-7}
3	1.6×10^{-4}	1.6×10^{-7}	1.6×10^{-10}
4	1.9×10^{-5}	1.9×10^{-9}	1.9×10^{-13}
5	2.9×10^{-6}	2.9×10^{-11}	2.9×10^{-15}

For this example we assumed that all intermediate mutants have the same selective disadvantage, s. The mutation rate is $\mu = 3 \times 10^{-5}$.

obtain the rate of convergence towards equilibrium for any of the mutant strains, but the analytical expressions are too complex to be shown here. However, it is interesting to compare the times until a 2- or more-error mutant attains a given frequency relative to the wild-type if (i) the intermediate mutants cannot replicate at all or (ii) all mutants have the same selective disadvantage. For example, a 2-error mutant reaches a frequency of 10^{-8} about five times faster under scenario (ii) than under scenario (i). Hence, the presence of intermediate mutants with positive growth rate not only affects the equilibrium frequency of a n-error mutant, but also its rate of appearance in a patient. Note, however, that the time necessary for a mutant to reach a given percentage of its equilibrium value is larger under scenario (ii), simply because the equilibrium value is much higher in this scenario.

9.4   Some practical implications

The success of drug therapy depends to a large extent on whether or not resistant virus is present in patients prior to treatment. Based on a quasispecies model of virus population dynamics, we derived analytical expressions for the pre-treatment frequency of resistant mutants. We calculated how the expected frequency depends on the number of point mutations between wild-type and mutant, the selective disadvantage of the mutant and all its intermediates, and the mutation rate. For example, for a mutation rate of 3×10^{-5} and a selective disadvantage for the mutant and all intermediates of 1%, we find that the frequency of a 3-error mutant relative to wild-type is about 10^{-7}. If there are about 10^8 productively infected cells in a patient (Chun *et al.* 1997*a*), then such a mutant may well be present before the start of therapy.

There are several practical conclusions. The probability of treatment failure due to viral resistance can be reduced in three ways:

(i) Treatment should start early when virus load is low and the frequency of resistant mutants is small (assuming that most patients get infected by drug-sensitive wild-type virus).

(ii) Treatment should commence immediately with multiple drugs (3 or more). It is clearly disadvantageous to put patients on one drug and add other drugs later, as this gives the virus the possibility to develop resistance to each drug sequentially.

(iii) Treatment should combine drugs for which resistance mutations are known to involve a considerable selective disadvantage in the absence of drug. It may be of great importance to identify such drug combinations, because mutants resistant to these combinations may not exist in patients prior to treatment.

9.5 Further reading

Zidovudine resistance of HIV-1 was first characterized by Larder *et al.* (1989). These findings were followed by many publications on HIV resistance. We can only mention a few examples: Schuurman *et al.* (1995), Eron *et al.* (1995), Condra *et al.* (1995), and Richman (1994*a*,*b*). Mansky and Temin (1995) measured the mutation rate of HIV-1 *in vivo*. Ribeiro *et al.* (1998) compute the frequency of resistant mutants prior to treatment.

10

EMERGENCE OF DRUG RESISTANCE

As we have discussed in the previous chapter, anti-viral treatment of HIV-1 infection often fails because of the emergence of resistant virus. In single-drug treatment the virus population in a patient can be completely resistant within a few weeks of initiating therapy. Such treatment usually leads to a rapid, but short-lived decline in plasma virus load; after a few weeks the plasma virus load is back to its original pre-treatment level. Similarly there is an increase in the CD4 cell population followed by a return to baseline.

Angela McLean, then at the University of Oxford, was among the first to model these dynamics. In the early 1990s she pointed out that the observed rebound of virus during therapy is a consequence of the increased abundance of CD4 cells. The inhibition of virus replication leads to an increase in the number of target cells, which provides new opportunities of viral infection. Such dynamics are reminiscent of predator–prey interactions in biology. Reducing the abundance of predators may cause an increase in the number of prey, which in turn causes predators to rise again. (The mathematical equations describing such interactions are nonlinear and lead to such oscillations.) In 1992, McLean and Nowak analysed models of viral resistance and pointed out that the rebound of virus during treatment can either contain wild-type or resistant mutant virus depending on their relative replication rates during treatment. Again, resistant virus emerges during treatment partly as a consequence of increased target cell abundance.

For simplicity, consider a situation with only two virus variants: the wild-type virus (which predominates the population before treatment) and a mutant virus. Denote by R_1 and R_2 the basic reproductive ratios of wild-type and mutant before therapy. By definition we have $R_1 > R_2$; the wild-type is fitter than the mutant otherwise it would not be wild-type.

Denote by R_1' and R_2' the basic reproductive ratios of wild-type and mutant during therapy. As outlined in Chapter 9, we consider a virus mutant as being resistant, if during treatment it has a basic reproductive ratio greater than one. This means that it cannot be eradicated by treatment; in the absence of competition with other virus mutants it would persist indefinitely. Conversely we consider a virus mutant sensitive to treatment, if its basic reproductive ratio during treatment is less than 1, which implies that—in principle—it can be eradicated by treatment.

There are four possible scenarios of anti-viral treatment (Fig 10.1):

(i) A very weak drug (low dose or low efficacy) may not reduce the rate of wild-type reproduction below mutant reproduction, i.e. $R_1' > R_2' > 1$. In this case, both wild-type and mutant virus are resistant to treatment; the mutant virus will not be selected. Nothing much will change.

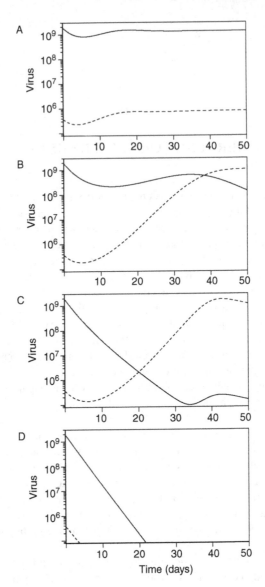

Fig. 10.1 Dynamics of drug treatment if (resistant) mutant virus is present prior to therapy. Before treatment, the basic reproductive ratios of wild-type and mutant virus are given by R_1 and R_2, respectively. Drug therapy reduces the basic reproductive ratios to R_1' and R_2'. There are four possibilities depending on dosage and efficacy of the drug: (A) if $R_1' > R_2'$, then mutant virus is still outcompeted by wild-type. Emergence of resistance will not be observed. Equilibrium virus abundance during treatment is similar to the pre-treatment level. (B) If $R_2' > R_1' > 1$, resistance will eventually develop, but the initial resurgence of virus can be due to wild-type. (C) If $R_2' > 1 > R_1'$, resistant virus rises rapidly. In (B) and (C) the exponential growth rate of resistant virus is approximately given by $a(R_2' - 1)$, thus providing an estimate for the basic reproductive ratio of resistant virus during treatment. (D) If $1 > R_1', R_2'$, then

(ii) A stronger drug may reverse the competition between wild-type and mutant such that $R_2' > R_1' > 1$. Here again both wild-type and mutant are resistant to treatment, but the mutant virus will eventually dominate the population after long-term treatment. The initial resurgence of virus can be mostly wild-type.

(iii) A still stronger drug may reduce the basic reproductive ratio of wild-type below one, $R_2' > 1 > R_1'$. Here the mutant, but not the wild-type is resistant to treatment. The wild-type virus should decline roughly exponentially after start of therapy and be maintained in the population at very low levels (only because of mutation).

(iv) In this happy case, a very effective drug may reduce the basic reproductive ratio of both wild-type and mutant below one, $1 > R_1 > R_2$. Thus, neither wild-type nor mutant are resistant to treatment. The virus population will be eliminated (in the absence of additional mutations).

Therefore, resurgence of (wild-type or mutant) virus during treatment is a consequence of an increasing abundance of target cells. There is experimental evidence that rising target cell levels can lead to a rebound of wild-type virus during zidovudine treatment. Preventing target cell increase could therefore maintain the virus at low levels (De Boer and Boucher 1996).

During 1996, Sebastian Bonhoeffer, then at the University of Oxford, used the basic model of virus dynamics to explore the emergence of drug resistance. He obtained three fundamental results, which we generously propose to call *Bonhoeffer's Laws of Anti-viral Treatment and Resistance*.

First Law (Return to Baseline): If treatment cannot eradicate the virus, then the virus load will eventually return back to baseline.

The first law states that according to the basic model of virus dynamics the equilibrium virus load should only be affected strongly by drug treatment if the drug shifts the virus population close to extinction. For less effective drugs the equilibrium virus load before and during long-term treatment is essentially the same, largely independent of the intrinsic growth rates of wild-type and resistant virus.

Second Law (Conservation of Area): The benefit of treatment is independent of the level of inhibition of sensitive virus. A stronger drug selects for faster emergence of resistant virus; the area under the virus load curve is constant.

The second law means that, provided drug resistant mutants exist prior to treatment, the total benefit of drug treatment (as measured by the total gain of CD4 cells or the total

---

both wild-type and mutant virus will decline exponentially. Parameter values: $\lambda = 10^7$, $d = 0.1$, $a = 0.5$, $u = 5$, $k_1 = k_2 = 500$, $\beta_1 = 5 \times 10^{-10}$, $\beta_2 = 2.5 \times 10^{-10}$. Hence, $R_1 = 10$ and $R_2 = 5$. Treatment reduced β_1 and β_2 such that: (A) $R_1' = 3$, $R_2' = 2.5$; (B) $R_1' = 1.5$, $R_2' = 2.25$; (C) $R_1' = 0.5$, $R_2' = 2$; (D) $R_1' = 0.1$, $R_2' = 0.5$; and the continuous line is wild-type virus, whereas the broken line denotes mutant.

reduction of virus load compared to baseline) is independent of the degree of inhibition of sensitive virus by the drug. The time necessary until a significant fraction of the virus population is resistant is shorter the more effective the drug.

Third Law (Probability of Resistance): The probability that resistant virus emerges during effective therapy is less than the probability that resistant virus is already present before therapy.

The third law compares the probabilities of pre-existence and emergence of drug resistant mutants. As a lower limit, the fraction of cells infected with resistant virus in drug-naive patients is given by the mutation rate of sensitive to resistant virus. Thus, even if drug resistant mutants are heavily selected against in the absence of drug, we expect prior to treatment on average one resistant infected cell in an infected cell population size of the reciprocal of the mutation rate from sensitive to resistant virus. We show that if the drug can eradicate the sensitive wild-type virus, then the probability that resistant mutants appear after the start of therapy is smaller than the probability that resistant virus is present before therapy. If the drug cannot eradicate the wild-type then mutants that are unlikely to pre-exist may take a long time to appear after the start of therapy.

Bonhoeffer's laws are direct consequences of the basic model of viral dynamics as introduced in Chapter 3. In the following we will show how they can be derived from the basic model. These results are dependent on certain assumptions of the basic model. We will emphasize where the results agree, and where they seem to be in conflict, with experimental evidence.

In Section 10.1 we very briefly review the basic model of virus dynamics. In Section 10.2 we extend the basic model to include equations for wild-type and mutant viruses, and derive and discuss the first and second law. In Section 10.3 we derive and discuss the third law, and in Section 10.4 we summarize the results.

10.1   The basic model

Let us recall the basic model of virus dynamics:

$$\frac{dx}{dt} = \lambda - dx - \beta x v, \tag{10.1}$$

$$\frac{dy}{dt} = \beta x v - ay, \tag{10.2}$$

$$\frac{dv}{dt} = ky - uv. \tag{10.3}$$

Here x, y, v denote, respectively, uninfected cells, infected cells and free virus. The basic reproductive ratio, R_0, is defined as the average number of secondary infected cells arising from one infected cell placed into an entirely susceptible cell population. The basic reproductive ratio is given by $R_0 = \lambda \beta k/(adu)$. If $R_0 < 1$, then the virus can neither establish nor maintain an infection. If $R_0 > 1$, then the system converges to the equilibrium $\hat{x} = au/(\beta k)$, $\hat{v} = \lambda k/(au) - d/\beta$, $\hat{y} = \lambda/a - du/(\beta k)$.

Implicit in this model are a number of assumptions that are worthy to point out. First, the model assumes that the contribution of the immune responses to the death of infected cells or free virus (and to reducing the rate of infection of new cells) is either negligible or constant over the time span of interest. While this assumption may be justified for the short-term virus dynamics following drug treatment, it need not hold for the long-term. Second, the dynamics of the susceptible cell population assumes a constant immigration of such cells from a pool of precursors. There are other possibilities, and the specific details could affect the long-term behaviour of the model. Third, the model neglects the loss of free virus due to the uptake by susceptible cells assuming that this loss is small compared to the natural decay of free virus (or remains constant). This assumption makes the viral decay independent of the size of the susceptible cell population.

Since the decay rate of free virus, u, is usually much larger than the decay rate of the virus-producing cell population, a, we may assume to a good approximation that free virus is in steady-state (i.e. $dv/dt = 0$), which implies that the free virus load is proportional to the infected cell load. In the following we will therefore eliminate v from eqn (10.3) by substituting $v = ky/u$ and $b = \beta k/u$ to obtain the much simpler pair of equations:

$$\frac{dx}{dt} = \lambda - dx - bxy, \tag{10.4}$$

$$\frac{dy}{dt} = bxy - ay. \tag{10.5}$$

10.2 Emergence of resistance during drug treatment

Assume that a small population of resistant virus pre-exists in the patient prior to drug treatment. We are interested in the rise of resistant mutant following drug therapy. Extending the basic model to distinguish between cells infected with resistant virus, y_r, and cells infected with sensitive virus, y_s, we get:

$$\frac{dx}{dt} = \lambda - dx - (sb_s y_s + b_r y_r)x, \tag{10.6}$$

$$\frac{dy_s}{dt} = (sb_s x - a)y_s, \tag{10.7}$$

$$\frac{dy_r}{dt} = (b_r x - a)y_r. \tag{10.8}$$

The parameters b_s and b_r reflect the overall replication rates of sensitive and resistant virus. Treatment is reflected by the parameter s (between 0 and 1), which describes the inhibition of infection of susceptible cells by sensitive virus. This is an accurate description of the effect of reverse transcriptase inhibitors (like AZT or 3TC), which impair reverse transcription of viral RNA into DNA and thereby prevent infection of susceptible cells. Since the turnover of free virus is fast, it is also a good approximation for protease inhibitors (such as ritonavir), which prevent infected cells from producing infectious virus particles.

For simplicity we assume that resistant virus is not affected by the drug. (We will point out below which conclusions are affected by this assumption.) As the resistance mutations are in the genes coding for the reverse transcriptase (RT) or the protease, sensitive and resistant virus may differ in their replication rates. In the absence of drug, the replication rate of the sensitive virus, b_s, is larger than the replication rate of the resistant virus, b_r. Otherwise the resistant virus would dominate the virus population prior to treatment. The death rate of infected cells is assumed not to be affected by the resistance mutation.

10.2.1   *Equilibrium properties*

There are two stable equilibria depending on the efficiency of drug treatment. If $s > b_r/b_s$ then the system converges to $\hat{x} = a/(sb_s)$, $\hat{y}_s = \lambda/a - d/(sb_s)$, $\hat{y}_r = 0$ and the virus population consists entirely of sensitive virus despite drug treatment. If $s < b_r/b_s$ then the system converges to $\hat{x}' = a/b_r$, $\hat{y}'_s = 0$, $\hat{y}'_r = \lambda/a - d/b_r$ and resistant virus will outcompete sensitive virus. Since we neglect mutation from sensitive to resistant virus (and vice versa) there is no coexistence of sensitive and resistant virus at equilibrium. A model including mutation will be discussed below.

Let us consider a sufficiently strong drug treatment such that $s < b_r/b_s$. Comparing the equilibria in absence and presence of treatment, we obtain for the factor of reduction of virus load

$$\alpha = \frac{1 - 1/R_0}{1 - 1/R'_0}, \tag{10.9}$$

where $R_0 = \lambda b_s/(ad)$ and $R'_0 = \lambda b_r/(ad)$ are the basic reproductive ratios before and during treatment, respectively. Put another way, to achieve an α-fold reduction in equilibrium virus load the basic reproductive ratio during treatment must be $R'_0 = (1 - 1/\alpha(1 - 1/R_0))^{-1} < (1 - 1/\alpha)^{-1}$, regardless of the basic reproductive ratio before treatment. Hence according to this model we only expect a large reduction in equilibrium virus load if the basic reproductive ratio during treatment is close to 1, which implies that the virus is at the verge of extinction.

This is difficult to reconcile with the sustained suppression of virus load observed in several drug trials. For example, patients receiving AZT–3TC combination therapy showed a 10–100-fold reduced equilibrium virus load, but in no patient is the virus ever maintained below detection limit for a longer period of time. This implies, that $R'_0 > 1$ in all patients. Given that R'_0 can in principle be any number smaller than R_0, it seems unreasonable to assume that R'_0 always falls exactly between 1 and $(1 - 1/\alpha)^{-1} \approx 1.1$. Therefore the basic model of virus dynamics cannot explain a sustained 10–100-fold reduction of equilibrium virus load as a consequence of drug therapy. In other words, the sustained suppression of virus load observed in patients treated with zidovudine–lamivudine combination therapy is not a trivial consequence of drug-induced reduction of virus infectivity.

The basic model of virus dynamics needs to be extended to explain these observations. Three extensions seem promising: (i) taking into account the dynamics of specific immune responses it is possible to obtain a reduced equilibrium virus load as a consequence of reduced viral infectivity. In this case, however, there should be a correlation between the strength of the anti-viral immune response in a patient and the amount of virus load reduction during treatment. In patients with a stronger immune response there will be a greater reduction in virus load. (ii) The drugs may have differential effects in different tissues or cell types. Suppose lamivudine–zidovudine treatment does drive the virus to extinction in a compartment which is mostly responsible for generating plasma virus load, but leaves the virus population essentially unaffected in a compartment which has only a small contribution on plasma virus load. In this case plasma virus load can be greatly reduced during long-term treatment. For a more detailed discussion and extensions to the basic model that may explain a strong sustained virus load see Bonhoeffer *et al.* (1997*a,b*).

10.2.2 Total gain of CD4 cells and total reduction of virus load are independent of inhibition of sensitive virus

Let us first define a criterion to assess the benefit of treatment. One such criterion would be the average time until a given fraction of the virus is resistant. Such a criterion is problematic, however, because the treatment strategy that maximizes the time for the appearance of resistance is not to treat at all. Therefore, we need to find another criterion which balances the value of preserving the drug's effectiveness with the value of reducing virus load or increasing CD4 cell count. Such a criterion is the total gain of uninfected CD4 cells or the total reduction of virus load over the duration of treatment compared to baseline before treatment (see Fig. 10.2).

In our model, the total gain of uninfected CD4 cells and the total reduction of virus load can be derived analytically. Let us assume that at the time $t = 0$ when therapy is started a small population of resistant virus, $y_r(0)$, pre-exists. Integrating eqn (10.8), we obtain for the total gain of susceptible cells after a time T (which shall be sufficiently long that the resistant virus attains its equilibrium, $\hat{y}'_r$, during treatment):

$$F(T) = \int_0^T (x(t) - \hat{x}) \, dt = 1/b_r (\ln(\hat{y}'_r/y_r(0))) + (\hat{x}' - \hat{x})T, \qquad (10.10)$$

where $\hat{x} = a/b_s$ and $\hat{x}' = a/b_r$ are the equilibrium densities of uninfected cells in absence and presence of treatment.

An analogous result can be obtained for the total reduction of virus load, $y = y_s + y_r$. Integration over $dx/dt + dy_s/dt + dy_r/dt$ yields:

$$x(T) - x(0) + y(T) - y(0) = \lambda T - d \int_0^T x \, dt - a \int_0^T y \, dt, \qquad (10.11)$$

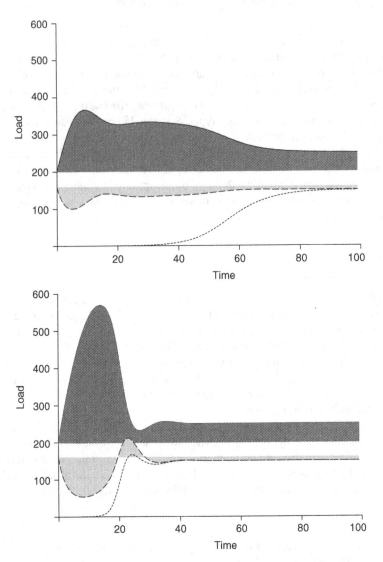

Fig. 10.2 Numerical simulation of drug treatment. The total benefit of treatment, as measured by the total gain of susceptible cells (dark shaded area, solid line) or the total reduction of virus load (light shaded area, dashed line) over a time long enough such that the resistant virus (dotted line) equilibrates during treatment, is independent of inhibition of sensitive virus by the drug (top: $s = 0.6$, bottom: $s = 0.4$). The shaded areas are of equal size in both plots. We assume, that before the start of therapy, the virus load is in equilibrium. When therapy is started a small fraction, f, of the total infected cells are assumed to be resistant. The other parameters (in arbitrary units) are: $\lambda = 100, d = 0.1, a = 0.5, b_s = 0.0025, b_r = 0.0020, f = 0.0001$.

where $x(0)$, $y(0)$, $x(T)$, and $y(T)$ are the uninfected cell and total infected cell load at the start of treatment and after a time T. Using $\hat{x} = a/b_s$ and $\hat{y} = \lambda/a - d/b_s$, we obtain

$$G(T) = \int_0^T (y(t) - \hat{y})\,dt = -d/a \int_0^T (x(t) - \hat{x})\,dt - \frac{x(T) - x(0) + y(T) - y(0)}{a},$$

(10.12)

where $\hat{y}$ is the total virus load before therapy.

Both, the total gain of susceptible cells, $F(T)$, and the total reduction of virus load, $G(T)$, depend logarithmically on the initial frequency of resistant virus. Therefore, the benefit of treatment depends only weakly on the frequency of resistant mutants prior to treatment. Interestingly, the benefit is independent of the inhibition of wild-type virus, s, by the drug (see Fig. 10.2). More precisely: b_r represents the replication rate of resistant virus in presence of the drug; if b_r is not affected by the drug, then, according to this model, the benefit of treatment is independent of the efficacy of the drug. But if we relax this assumption and assume that the infectivity of resistant virus in presence of drug is a function $b_r(s)$ of drug concentration, s, then the total benefit of treatment depends on s. Therefore, the total gain of CD4 cells and the total reduction of virus load depend only on the effect of the drug on the resistant virus but not on the sensitive virus.

The frequency of resistant mutants prior to treatment and their infectivity are likely to vary for different drugs. Hence, the model does not predict that the total gain in CD4 cells or the reduction of virus load should be the same for different drugs.

10.2.3 A stronger drug selects for faster emergence of resistance

The time necessary until a significant fraction of the virus is resistant depends on the inhibition of sensitive virus by the drug. Figure 10.3A shows that the stronger the inhibition, the faster the emergence of resistance. An approximate solution for the time until 50% of the total infected cell load is resistant virus is $T_{1/2} = (sb_s)/(ab_r)\ln([1 - b_r/b_s) b_r y_r(0)/d]$. Numerical simulation supports this analytic result, showing that $T_{1/2}$ depends strongly on s, but depends only logarithmically on the initial frequency of resistant infected cells prior to treatment. See Fig. 10.3B and C. The effect that resistant virus emerges faster the stronger the inhibition by the drug is based on the pre-existence of resistant virus prior to treatment. In the next section, we will show that if resistant virus does not pre-exist, then the time until resistant virus appears is longer the stronger the drug.

10.3 The probability of producing a resistant mutant during therapy

Let us now consider a situation where no resistant mutant is present at the time when therapy starts. What is the likelihood that a resistant mutant appears during therapy?

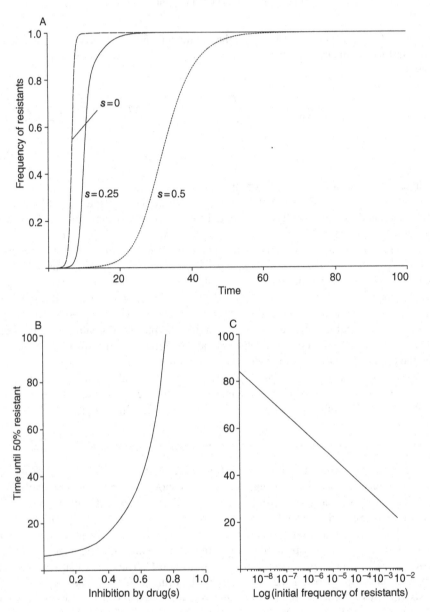

Fig. 10.3 The stronger the treatment the faster the evolution of resistance. Plot A shows the frequency of resistant virus for three different values for the inhibition of sensitive virus by the drug. Plots B and C show the time, $T_{1/2}$, necessary until 50% of the total cells are infected with resistant virus as a function of the inhibition of sensitive virus, and the initial frequency of resistant virus prior to treatment. Parameters (in arbitrary units): $b_s = 0.005$, $b_r = 0.004$, $f = 0.0001$ (plots A and B), and $s = 0.6$ (plot C), all other parameters as in Fig. 10.1.

In absence of resistant virus the dynamics of the virus population during drug treatment are:

$$\frac{dx}{dt} = \lambda - dx - sb_s x y_s,$$ (10.13)

$$\frac{dy_s}{dt} = sb_s x y_s - a y_s.$$ (10.14)

The total number of cells infected after the start of therapy is given by $sb_s \int_0^\infty x y_s$. We do not have exact analytical solutions for $x(t)$ and $y_s(t)$, but we can approximate the integral (see Fig. 10.4).

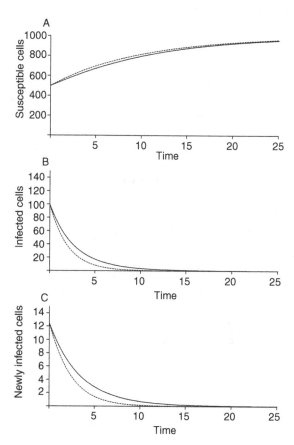

Fig. 10.4 Approximation of the new infections after therapy is started. The solid line represents the numerical solution of the full model and the dotted line is the approximation described in the main text. Plots A and B show the dynamics of the susceptible and infected cell population, respectively. Plot C shows the total production rate of infected cells during treatment given by $sb_s y_s x$. Parameters: $s = 0.25$, $b_s = 0.001$, all other parameters as in Fig. 10.1.

We distinguish between two scenarios: (i) the drug treatment is sufficiently potent to eradicate the wild-type virus (provided no resistant mutant emerges) and (ii) the drug treatment cannot eradicate the wild-type virus but reduces its replication rate.

The first scenario implies that the basic reproductive ratio during treatment, $R_0' = s\lambda b_s/(ad)$, is smaller than unity. If we assume that the basic reproductive ratio before treatment is distinctly larger than unity, then the total natural death rate of infectable cells, dx, far outweighs the total rate of infection of cells, $sb_s x y_s$. Hence, we get for the infectable cell population $x(t) \approx \lambda/d - (\lambda/d - x_0)\,\mathrm{e}^{-dt}$, where $x_0(= a/b_s)$ is the initial load of infectable cells when treatment is started. We can also neglect $sb_s x y_s$ in comparison to ay_s, and we get for the dynamics of the infected cell population $y_s(t) \approx y_0\mathrm{e}^{-at}$, where y_0 is the infected cell load when therapy is started. Integrating over $sb_s x y_s$ we get for the total number of infected cells produced after the start of therapy:

$$N \approx R_0'\left[1 - \left(1 - \frac{1}{R_0}\right)\frac{a}{a+d}\right]y_0.$$

Obviously, when the inhibition is complete ($R_0' = 0$), then we have $N = 0$. For incomplete inhibition ($1 > R_0' > 0$) we find that the number of infected cells produced during treatment, N, is less than the number of infected cells, y_0, prior to treatment (since $R_0' < 1$ and $R_0 > 1$). Thus, if a resistant mutant is unlikely to pre-exist in untreated patients, it is also unlikely to appear during drug treatment, provided that the treatment can eradicate the sensitive wild-type virus.

If the drug cannot eradicate the sensitive virus, the resistant mutant will eventually appear, given enough time. Neglecting the transients before the infected cell load reaches its reduced equilibrium, $\hat{y}_s' = \lambda/a - d/(sb_s)$, during treatment, we can obtain a rough estimate for time until the first resistant mutant is produced. In equilibrium, the total rate of infection of susceptible cells equals the total death rate of infected cells. Thus, the total daily production of infected cells is given by $a\hat{y}_s'$. We expect one resistant mutant to be produced in $1/\mu$ infections. Hence we get for the average time until a resistant mutant is produced $T = 1/(\mu a\hat{y}_s')$. If the equilibrium virus load during treatment is reduced by a factor α compared to baseline (i.e. $\hat{y}_s = \alpha\hat{y}_s'$), we get $T \approx \alpha/(\mu a\hat{y}_s)$. Hence if a mutant is unlikely to pre-exist (i.e. $\mu\hat{y}_s \ll 1$), then it will take a long time, on average, until the first resistant mutant is produced.

10.4 Summary

If resistant virus exists prior to treatment, we find that the larger the inhibition of sensitive virus by the drug, the shorter the time until a significant fraction of the virus is resistant. However, the total benefit of treatment (as measured by the total gain of susceptible cells or the total reduction of virus load compared to baseline before treatment) is independent of the degree of inhibition of the sensitive virus by the drug. Provided that the resistant virus is not affected by the drug, the basic model of virus dynamics predicts that the total benefit of treatment is independent of drug dosage. However, the total benefit depends on the frequency of resistant virus prior to treatment, which in turn is likely to be specific for each drug. Therefore, the total gain of CD4 cells and the reduction of virus load should vary between different drugs. It should also vary for different drug dosages, if the infectivity of resistant virus is strongly affected by drug dosage.

If resistant virus does not pre-exist, we distinguish between two cases. If the drug is sufficiently potent to eradicate the sensitive virus, then a mutant that is unlikely to exist prior to therapy is also unlikely to be produced during therapy. If the drug cannot eradicate the wild-type, then after sufficient time resistant mutants will appear. However, mutants that are unlikely to pre-exist will take a long time to appear.

Taken together these results suggest that viral resistance is a consequence of (i) either the drug not being sufficiently potent to eradicate the virus or (ii) the presence of resistant mutants prior to treatment. In the latter case we expect resistance to appear rapidly after the start of therapy. In the former case resistant mutants will appear slowly, especially if wild-type and mutant virus have similar basic reproductive ratios during treatment.

10.5   Further reading

Mathematical models that deal with the dynamics of anti-viral treatment and the emergence of virus resistance are McLean *et al.* (1991), McLean and Nowak (1992*a,b*), Frost and McLean (1994), Stilianakis *et al.* (1997), De Boer and Boucher (1996), Goudsmit *et al.* (1996) and Nowak *et al.* (1991). 'Bonhoeffer's Laws' are formulated in Bonhoeffer and Nowak (1997). Bonhoeffer *et al.* (1997*b*) explore the effect of anti-viral treatment on virus load. A review of virus dynamics and the evolution of resistance is Bonhoeffer *et al.* (1997*a*).

11

TIMING THE EMERGENCE OF RESISTANCE

In this chapter we describe the detailed dynamics of the rise of resistant virus during (single)-drug therapy. We assume that resistant mutants pre-exist prior to treatment and calculate the rate at which their abundance increases during treatment. We provide analytic approximations for the fall of wild-type virus and the rise of resistant virus. We calculate the time it takes until the resistant mutant reaches 50% prevalence in free virus and infected cell populations. The calculations in this chapter are based on the basic model of virus dynamics; the underlying central assumption is that resistant virus rises as a consequence of an increase in target cell numbers.

We also discuss data from patients who received single-drug therapy (nevirapine) and show how they can be used together with the model to estimate demographic parameters of virus population dynamics. More than 90% of infected peripheral blood mononuclear cells (PBMC) harbour defective viral genomes. The half-life of these cells is about 100 days. Cells with replication competent virus are either actively engaged in virion production (and have a half-life of about 2 days) or seem latently infected with a half-life of about 10–20 days. There is a 20–40% probability that infection of a cell results in a replication competent provirus.

In Section 11.1 we outline the mathematical model, and in Section 11.2 we discuss the clinical data. Section 11.3 provides a short summary.

11.1 Theory

In HIV-1 infection there is rapid development of resistant virus to all known drugs. Often a single-point mutation can greatly reduce sensitivity to a particular drug. In this section we calculate the rate at which the abundance of a resistant mutant rises. Before drug therapy the resistant mutant virus is held in a mutation–selection balance. It is continuously generated by the sensitive wild-type, but has a slight selective disadvantage. (If it had a selective advantage it would dominate the virus population before therapy is given; if it was completely equivalent to the wild-type then its initial frequency would be subject to neutral drift.)

An appropriate system that describes this mutation–selection process is the following:

$$\begin{aligned}
\dot{x} &= \lambda - dx - \beta xv - \beta_m x v_m, \\
\dot{y} &= \beta(1 - \epsilon)xv - ay, \\
\dot{v} &= ky - uv, \\
\dot{y}_m &= \beta\epsilon xv + \beta_m x v_m - ay_m, \\
\dot{v}_m &= k_m y_m - uv_m.
\end{aligned} \tag{11.1}$$

Here x denotes uninfected cells; y and y_m denote cells infected by wild-type and mutant virus, whereas v and v_m denote wild-type and mutant virions, respectively. The parameter ϵ denotes the probability of mutation from wild-type to resistant mutant during one round of replication; ϵ will be within 10^{-3} to 10^{-5}. We can neglect back-mutation from mutant to wild-type. We assume that wild-type and mutant may differ in their rates at which they infect new cells, β and β_m, and the rates at which infected cells produce virus, k and k_m.

The basic reproductive ratios of wild-type and mutant virus are given by

$$R_0 = \frac{\beta \lambda k}{adu}, \tag{11.2}$$

$$R_m = \frac{\beta_m \lambda k_m}{adu}. \tag{11.3}$$

In the absence of drug treatment the wild-type has a higher basic reproductive ratio (i.e. fitness) which implies $\beta k > \beta_m k_m$. This condition specifies that selection favours wild-type virus. For small ϵ the steady-state of (11.1) is

$$x^* = \frac{au}{\beta k},$$

$$v^* = \frac{\lambda k}{au} - \frac{d}{\beta},$$

$$y^* = \frac{\lambda}{a} - \frac{du}{\beta k}, \tag{11.4}$$

$$y_m^* = y^* \epsilon \Big/ \left(1 - \frac{\beta_m k_m}{\beta k}\right),$$

$$v_m^* = \frac{k_m}{k} v^* \epsilon \Big/ \left(1 - \frac{\beta_m k_m}{\beta k}\right).$$

Let us now assume that the drug completely inhibits the replication of wild-type virus ($\beta = 0$), but does not affect the mutant. (The second assumption can easily be replaced by introducing a somewhat reduced growth rate of the mutant under drug therapy; this does not change the essentials of the model.) This gives rise to the equations:

$$\begin{aligned}
\dot{x} &= \lambda - dx - \beta_m x v_m, \\
\dot{y} &= -ay, \\
\dot{v} &= ky - uv, \\
\dot{y}_m &= \beta_m x v_m - a y_m, \\
\dot{v}_m &= k_m y_m - u v_m.
\end{aligned} \tag{11.5}$$

We want to solve this system subject to the initial conditions (11.1), the steady-state before drug treatment. Wild-type virus and infected cells decline as

$$y(t) = y^* e^{-at}, \tag{11.6}$$

$$v(t) = v^* \frac{u e^{-at} - a e^{-ut}}{u - a}. \tag{11.7}$$

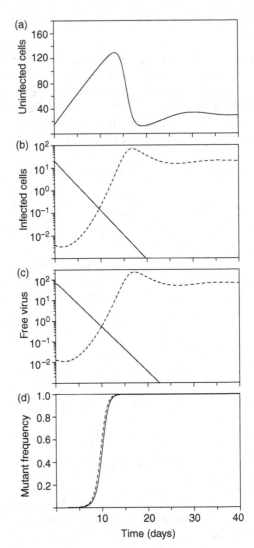

Fig. 11.1 A computer simulation of the basic model (eqn (11.1)) describing the rise of resistant mutant virus during drug therapy. Uninfected cells rise initially almost linearly with time. Wild-type virus declines exponentially with rate a both in terms of infected cells and free virus. Resistant mutant initially declines compared to its pre-treatment steady-state, but then rises rapidly as a consequence of an increase in uninfected cell numbers. Subsequently uninfected cells decline and the system converges to a new steady-state with the virus population consisting entirely of resistant mutant. Note that new and old steady-state have essentially equivalent x, y and v values, even if the replication rate of the resistant virus during drug therapy is much lower than the replication rate of wild-type virus before therapy. Parameter values are $\lambda = 10$, $d = 0.01$, $a = 0.5$, $u = 3$, $\beta = 0.01$, $\beta_m = 0.005$, $\epsilon = 0.0001$, $k = k_m = 10$ per day. This implies that the average life-times of uninfected cells, infected cells and free virus are, respectively, 100, 2 and 0.333 days. The basic

If the decay rate of free virus is much larger than the decay rates of infected and uninfected cells ($u \gg a, d$), then to a good approximation we may assume that $v_m = (k_m/u)y_m$ and the equations for the rise of resistant mutant become

$$\dot{x} = \lambda - dx - ds\frac{\beta_m k_m}{u}xy_m,$$
$$\dot{y}_m = y_m\left(\frac{\beta_m k_m}{u}x - a\right).$$

(11.8)

We rescale $\hat{x} = (d/\lambda)x$, $\hat{y}_m = y_m\beta_m k_m/(du)$ and obtain

$$\dot{\hat{x}} = d[1 - \hat{x}(1 + \hat{y}_m)],$$
$$\dot{\hat{y}}_m = a\hat{y}_m[R_m\hat{x} - 1].$$

(11.9)

It is useful to define

$$f = \frac{\beta_m k_m}{\beta k} = \frac{R_m}{R_0}.$$

(11.10)

The rescaled initial conditions are now

$$\hat{x}(0) = \frac{1}{R_0}, \quad \hat{y}(0) = \hat{\epsilon}.$$

(11.11)

Here $\hat{\epsilon}$ is defined (neglecting terms of relative order ϵ) as

$$\hat{\epsilon} = \epsilon R_m\left(1 - \frac{1}{R_0}\right)\bigg/(1 - f).$$

(11.12)

The solutions of system (11.9) follow along the lines of a similar mathematical system for the spread of infectious diseases in populations as analysed in Appendix C of Anderson and May (1991). There are two phases for the rise in resistant mutants (Fig. 11.1). As long as $\hat{y}_m$ is small compared to 1, the system is approximately

$$\dot{\hat{x}} = d(1 - \hat{x}),$$
$$\dot{\hat{y}}_m = a\hat{y}_m(R_m\hat{x} - 1).$$

(11.13)

---

reproductive ratio of wild-type virus before treatment is $R_0 = 66.6$ and of resistant virus during treatment is $R_m = 33.3$. In the simulation it takes about $t_y = 9.7$ days for the resistant virus to increase to a relative frequency of 50% in the infected cell population. Using eqn (11.23) of our analytical approximation, we obtain $t_y = 10.1$ days; the exact solution of the quadratic eqn (11.21) suggests $t_y = 8.8$ days. (a) Uninfected cells, x versus time. (b) Infected cells: wild-type virus, y, continuous line; mutant virus, y_m, broken line. (c) Free virus population: wild-type virus, v, continuous line; mutant virus, v_m, broken line. (d) Relative frequency of mutant virus in free virus (continuous line) and infected cell population (broken line).

We have

$$\hat{x}(t) = 1 - (1 - 1/R_0)e^{-dt}, \tag{11.14}$$

$$\hat{y}_m(t) = \hat{\epsilon} \, \exp\left[a\left\{(R_m - 1)t - \frac{R_m}{d}\left(1 - \frac{1}{R_0}\right)(1 - e^{-dt})\right\}\right]. \tag{11.15}$$

Interestingly, the resistant mutant, $\hat{y}_m(t)$, falls initially, because its abundance is at first still roughly set by the equilibrium in a mutation–selection balance. As the mutational influx from the wild-type disappears, because the drug stops wild-type replication, the mutant falls, because the uninfected cell population is at a level too low to maintain the mutant. But eventually the rise in the uninfected cell population, $\hat{x}(t)$, causes the resistant mutant to increase (once $R_m\hat{x} > 1$). Note that resistant mutant virus grows only because the uninfected cell population increases. Phase I ends when $\hat{y}_m(t)$ becomes of order unity, which for small f is (very) approximately at $t = 2/(R_m d)$.

In phase II there is fast exponential growth of $\hat{y}_m(t)$, essentially on a $1/a$ time-scale. Correspondingly $\hat{x}(t)$ will fall, and $\hat{y}_m(t)$ will peak when $\dot{\hat{y}}_m = 0$ or $\hat{x}(t) = 1/R_m$. At this point, $\hat{x}(t)$ is still decreasing, so $\hat{y}_m(t)$ will now fall (still on the time-scale of $1/a$ at first), until $\hat{x}(t)$ ceases to decline when $\hat{x} = 1/(1 + \hat{y}_m)$. At this point $\hat{y}_m$ will be enough below unity that $\hat{x}$ climbs again, but $R_m\hat{x} < 1$ and $\hat{y}_m$ continues to fall. Eventually (time-scale $1/d$) $\hat{x}(t)$ will climb back to $R_m\hat{x} > 1$, whereafter $\hat{y}_m(t)$ will rise and the cycle repeats. There are slowly damped oscillations eventually leading to the equilibrium $\hat{x} = 1/R_m$ and $\hat{y}_m = R_m - 1$. The damping time-scale is of the order of $\tau_d = 2/(R_m d)$; in the later stages there are low amplitude oscillations of period $\tau_{osc} = 2\pi/\sqrt{ad(R_m - 1)}$.

In order to extract information about the 'demographic' (birth and death) parameters in eqn (11.1) from the available data, next we calculate the rise of resistant mutant in terms of the relative proportions of mutants among the populations of infected cells or free virus. The proportions of mutants in both these categories will rise, converging to essentially 100% during phase I of the dynamics. If R_m is significantly larger than unity, so that $dt < 1$ throughout phase I, eqn (11.15) can be Taylor-expanded to give

$$\hat{y}_m(t) = \hat{\epsilon} \, \exp[-(1 - f)at + \tfrac{1}{2}(R_m - f)adt^2 + \cdots]. \tag{11.16}$$

Let us define $\phi(t) = y_m(t)/y(t)$, where y_m and y now refer to the absolute values, before any rescaling. We have

$$\phi(t) = \frac{y_m(0)}{y(0)} \exp[f \, at + \tfrac{1}{2}(R_m - f)adt^2 + \cdots]. \tag{11.17}$$

Remember that $y_m(0)/y(0) = \epsilon/(1 - f)$. The relative proportion of mutant virus among the infected cells is then

$$Y(t) = \frac{y_m(t)}{y_m(t) + y(t)} = \frac{\phi(t)}{\phi(t) + 1}. \tag{11.18}$$

Similarly, the fraction of mutant in the free virus population is

$$V(t) = \frac{v_m(t)}{v_m(t) + v(t)} = \frac{\phi(t)}{\phi(t) + (k/k_m)}. \tag{11.19}$$

If $k = k_m$, which means that wild-type and mutant do not differ in the rate at which infected cells produce virus, then the rise of mutant in the free virus and infected cell populations occurs simultaneously. If $k > k_m$ then the mutant rises first in the infected cell population. If $k < k_m$ (which is possible but unlikely, because the wild-type is expected to have a higher fitness) then the mutant rises first in the free virus population.

In particular, the time at which the mutant virus constitutes 50% of all infected cells, t_y, is given by the solution of

$$\phi(t_y) = 1, \tag{11.20}$$

with $\phi(t)$ defined by eqn (11.17). Under the approximation of eqn (11.17), t_y is given by the solution of the quadratic equation

$$\tfrac{1}{2}(R_m - f)adt_y^2 + f\,at_y - \ln[(1-f)/\epsilon] = 0. \tag{11.21}$$

For specified values of the parameters R_m, f, a, d, and ϵ, the solution is routine. In the limit when R_m is significantly greater than f, a significantly larger than d (infected cells live less long than uninfected ones), and $\epsilon \ll 1$, we have the approximate solution

$$t_y \approx \left(\frac{1}{fa}\right)\ln\left[\frac{1-f}{\epsilon}\right], \quad \text{if } \frac{2f^2a}{R_md} > \ln\left(\frac{1-f}{\epsilon}\right), \tag{11.22}$$

$$t_y \approx \sqrt{\frac{2}{R_mad}\ln\left[\frac{1-f}{\epsilon}\right]}, \quad \text{if } \frac{2f^2a}{R_md} < \ln\left(\frac{1-f}{\epsilon}\right). \tag{11.23}$$

Likewise, the time at which the mutant virus represents 50% of all free virus, t_v, is given by $\phi(t_v) = k/k_m$. The approximate results of eqns (11.22) and (11.23) again pertain, except now ϵ is replaced by $\epsilon k_m/k$ in the argument of the logarithmic term.

Finally, the difference between the times when the mutant virus constitutes 50% of all infected cells, and 50% of all free virus, $\Delta = t_y - t_v$, is given under the approximation of the quadratic equation (11.21) for t_y, and its analogue for t_v, as:

$$\Delta = \frac{\ln(k_m/k)}{fa + 1/2(R_m - f)ad(t_y + t_v)}. \tag{11.24}$$

Thus, empirical observations of t_y and t_v enable us to make inferences about the rate constants. In particular, eqn (11.21) or (11.23) may be used to obtain a rough estimate on the basic reproductive ratio, R_m, of the mutant under drug therapy. It must be kept in mind that this analysis is based on the assumption that t_y and t_v are both attained during 'phase I' of the mutant virus' dynamics (i.e., with $\hat{y}_m(t) < 1$); in any applications, either to understanding simulations or analysing real data, the validity of this approximation should be checked for consistency, once inferences have been drawn.

Figure 11.1 shows the dynamics of the emergence of drug-resistant virus in the basic model. In the simulation it takes about 9.7 days for the resistant virus to reach 50% prevalence in the infected cell population. This agrees well with a prediction of 10.1 days from the analytic approximation given by eqn (11.23).

In Appendix A, we discuss an extended model that describes the rise of resistant virus in long-lived infected cells.

11.2   Observation

Three patients were treated with the reverse transcriptase inhibitor nevirapine (NVP) by George Shaw and his group (Nowak *et al.* 1997). Plasma virus load, CD4 cell counts and infected PBMC were measured sequentially after the start of therapy. The amount of proviral DNA was measured in peripheral blood mononuclear cells (PBMC), which gives the total proportion of HIV-infected PBMC. Using a limiting dilution assay for quantifying infectious units per 10^6 PBMC the frequency of cells that harbour replication competent virus (either latently or actively replicating) was determined. In all three compartments (plasma virus, infectious PBMC and total infected PBMC) the proportions of NVP-sensitive wild-type virus and NVP-resistant mutant virus at day 0, 14, 28, 42 and 140 after initiating therapy were quantified (Fig. 11.2).

Table 11.1 gives information on viral load in terms of free virus and infected cells in 1 ml of blood in all three patients before start of therapy. Note that most HIV-infected PBMC appear to harbour replication defective virus; there are only between 15 and 30 infectious units within PBMC in 1 ml blood, but between 420 and 550 PBMC carry HIV provirus. We have to bear in mind, however, that most HIV-infected cells and perhaps also most cells that produce plasma virus are in the lymph system and not in the peripheral blood.

After start of therapy, plasma virus load declines rapidly within the first 2 weeks and subsequently increases with the emergence of resistant virus. In patients 1625 and 1605 the initial decay of wild-type plasma virus between days 0 and 14 occurs with a half-life of 2.6 and 2.1 days, respectively. Over the same time infected PBMC that harbour replication competent wild-type virus decay only with half-lives of, respectively, 4.8 and 3.0 days. This suggests that infected PBMC contain a subset of cells with a longer half-life than those cells which produce the majority of plasma virus. Between days 14 and 42, infected PBMC with replication competent wild-type virus decline with half-lives of 9.7 days in patient 1625 and 21 days in patient 1605. If we assume that by day 14 actively infected cells with wild-type virus (half-life of about 2 days) have essentially disappeared, we can use these estimates to extrapolate backwards in time and estimate what fraction of cells were latently infected at day 0. For patient 1625 we obtain that about 40% of PBMC with replication competent virus were latently infected, which is about 7 in 10^6 PBMC. For patient 1605 we obtain 7% which comes to about 1 in 10^6 PBMC. For patient 1624 these calculations are not possible because we lack measurements for day 14. Table 11.2 shows the relative abundance of total infected cells, actively infected cells and latently infected cells in the PBMC population.

NVP-resistant virus rises rapidly in all three patients in the free plasma virus population followed with a small delay in infected cells harbouring replication competent virus and with considerable delay in infected cells harbouring HIV provirus (Fig. 11.2). At day 28 plasma virus contains 100% resistant virus in patient 1625, 76% in patient 1605 and 92% in patient 1624. It takes between 10 and 20 days for the resistant virus to reach 50% prevalence in the free virus population.

Using eqn (11.23) we can get a very crude estimate for the basic reproductive ratio of the resistant mutant under drug therapy. From eqn (11.23) we obtain $R_m = \alpha/t_y^2$ with $\alpha = (2/ad)\ln[(1-f)/\epsilon]$. Assuming $a = 0.5, d = 0.01, \epsilon = 0.0001$ and f somewhere between 0 and 0.99 we get $\alpha \approx 2000 - 4000$. If we take the rise of resistant mutant in the free virus population as indicative of the rise in the actively virus producing infected cell population, i.e. $t_y \approx t_v$ (which holds if $k \approx k_m$), then an experimental observation of $t_v \approx 10$ days suggests a basic reproductive ratio of $R_m \approx 20-40$. Thus the intrinsic growth rate of the NVP-resistant mutant virus if target cell availability was not limiting may be between 20 and 40 infected cells arising from any one infected cell. This estimate is of course only very crude.

NVP-resistant mutant rises only slowly in the DNA provirus population. After 140 days between 65% and 75% of infected PBMC harbour mutant provirus, which suggests that the turnover rate of infected PBMC at large is slow, with a half-life of roughly 100 days. The time lag between viral variants that emerge in the plasma RNA population and the proviral DNA population has also previously been noted by studies of HIV evolution in single patients (Simmonds *et al.* 1991).

11.2.1   *The probability of producing replication competent provirus*

We can also obtain a very crude estimate of the probability that infection of a cell will result in defective or replication competent provirus. Consider the following model

$$\begin{aligned}
\dot{x} &= \lambda - dx - \beta xv, \\
\dot{y}_1 &= q_1\beta xv - a_1 y_1, \\
\dot{y}_2 &= q_2\beta xv - a_2 y_2, \\
\dot{v} &= ky_1 - uv.
\end{aligned} \tag{11.25}$$

Here y_1 are infected cells that harbour a replication competent virus genome, whereas y_2 are infected cells with a defective virus genome. The death rate of a y_1 cell is a_1, and the death rate of a y_2 cell is a_2. We expect $a_1 > a_2$. Free virus is only produced by y_1 cells. Infection of a cell either results in a cell with replication competent provirus (this happens with probability q_1) or in a cell with a defective provirus (with probability q_2). Clearly $q_1 + q_2 = 1$.

At equilibrium, we have the relation

$$\frac{a_1 y_1}{q_1} = \frac{a_2 y_2}{q_2}. \tag{11.26}$$

The experimental data provide direct information on y_1 and y_2 before start of therapy. We estimate a_1 from the decline of infected PBMC harbouring replication competent

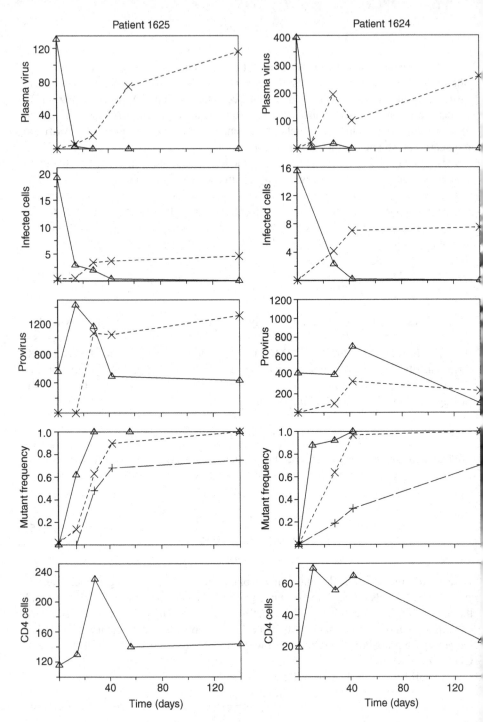

Fig. 11.2 Experimental data from three patients treated with the anti-HIV drug nevi-rapine. 'Plasma virus' denotes free virus particles in 1 μl plasma. 'Infectious cells' denotes PBMC harbouring replication competent HIV (per 1 ml blood). 'Provirus'

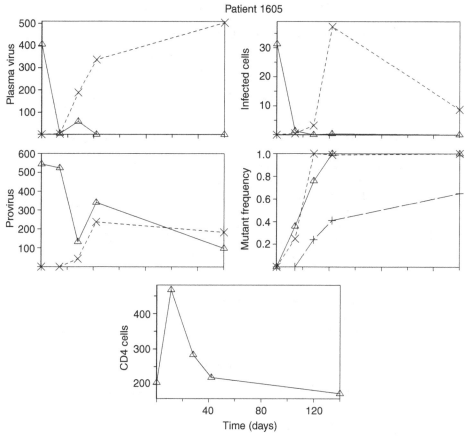

Patient 1605

denotes number of PBMC with HIV proviral DNA (per 1 ml blood). The continuous line is drug-sensitive wild-type virus, whereas the broken line denotes drug-resistant mutant virus. 'Mutant frequency' shows the rise of NVP-resistant mutant in the free virus population (continuous line), in the infectious PBMC population (broken line short dashes) and in the total infected PBMC population (broken line long dashes). In patients 1625 and 1624 resistant mutant rises first in the plasma population and then in the infected cell population (that has infectious virus). In patient 1605 the rise appears to be more or less simultaneous. In all three patients it takes much longer for the resistant mutant to establish itself in the provirus population. CD4 cell counts are shown per 1 μl blood.

wild-type virus and a_2 from the decline of defectively infected PBMC. Thus we obtain q_1 and q_2.

In all patients a_2 is approximately 0.0086 per day which is equivalent to a half-life of 80 days. For patient 1625 we have $y_1 = 17$, $y_2 = 463$ and calculate $a_1 = 0.15$ per day.

Table 11.1 *Pre-treatment data from three patients*

		1 ml blood contains			
			Infected PBMC		
				Infectious	
Patient	CD4 cells	Total PBMC	Provirus	virus	Plasma virus
1625	116 000	1 150 000	550	20	131 000
1605	205 000	1 750 000	540	31	407 000
1624	19 000	526 000	420	16	402 000

The table shows CD4 cells peripheral blood mononuclear cells (PBMC) and virus population size in terms of infected cells and free virus. 'Infected PBMC (provirus)' denotes cells that harbour HIV DNA in their genome. 'Infected PBMC (infectious virus)' denotes cells that contain replication competent HIV. Note that the large majority of infected cells (more than 90%) appears to be unable to produce infectious virus upon stimulation *in vitro*. 'Plasma virus' denotes free virus particles.

Table 11.2 *Relative amounts of HIV-infected PBMC in three patients before drug treatment*

	One million PBMC contain		
	Infected PBMC		
Patient	Provirus	Infectious virus	Latent virus
1625	480	17	7
1605	310	18	1
1624	800	29	N/A

'Infected PBMC (provirus)' denotes cells that harbour HIV DNA in their genome. 'Infected PBMC (infectious virus)' denotes cells that contain replication competent HIV, which can either be active or latent. 'Infected PBMC (latent virus)' denote cells that harbour replication competent, but latent virus. This number is estimated and can only be seen as a very rough guide (see main text). The estimate of latently infected cells is not possible for patient 1624.

This leads to $q_1 = 0.2$ and $q_2 = 0.8$. For patient 1605 we have $y_1 = 18$, $y_2 = 292$ and calculate $a_1 = 0.23$, leading to $q_1 = 0.22$ and $q_2 = 0.78$. For patient 1624 this analysis gives $q_1 = 0.37$. Therefore, it seems that if infection of a cell leads to a provirus the chance is between 0.2 and 0.4 that this provirus is replication competent. The fact that the large majority of infected PBMC harbour defective provirus is mostly a consequence of their slow turnover rate.

11.3 Summary

We have developed a mathematical framework for studying the emergence of resistant mutant virus during drug treatment. We assume that mutant virus exists already before start of therapy and is held in a mutation–selection equilibrium. Without drug the fitness of wild-type is higher than the fitness of mutant, but continuous mutation of wild-type virus (together with replication of mutant virus) produces a small amount of mutant virus.

Our analytic approximation for the emergence of resistance rests on the assumption that resistant virus rises as a consequence of increased abundance of target cells. From eqn (11.16) we have

$$v_m(t) \approx v_m(0) \exp[-(1 - R_m/R_0)at + \tfrac{1}{2}R_m(1 - 1/R_0)adt^2 + \cdots]. \qquad (11.27)$$

Here $v_m(t)$ is the abundance of mutant virus at time t after start of therapy, while a and d denote death rates of infected and uninfected cells. Thus, monitoring the rise of resistant mutant during drug therapy provides information on the basic reproductive ratio of HIV during infection. Note that this equation simply describes the rise of an already pre-existing mutant virus. If several mutations are necessary to confer resistance and if this involves complex evolutionary changes in the virus genome, then eqn (11.27) is not valid.

Our mathematical model in conjunction with data from three patients treated with the anti-HIV drug nevirapine also provides insights into the *in vivo* kinetics of HIV turnover: most of plasma virus is produced by cells with a half-life of about 1.5–2 days; latently infected cells have a half-life of about 10–20 days; most infected PBMC harbour replication defective provirus and have a half-life of about 100 days.

11.4 Further reading

Prolonged treatment with a single anti-HIV drug almost always results in the emergence of resistant virus (Larder *et al.* 1989, 1993, 1995; Larder and Kemp 1989; Richman 1990, 1994; Richman *et al.* 1994; St. Clair *et al.* 1991; Ho *et al.* 1994; Loveday *et al.* 1995; Markowitz *et al.* 1995). In AZT treatment, a number of mutations have been described that render the virus resistant against the drug (Boucher *et al.* 1990; 1992*a,b*; Mohri *et al.* 1993). In NVP and 3TC treatment, single point mutations appear to confer high level resistance against the drug (Richman *et al.* 1994; Schuurman *et al.* 1995; Wei *et al.* 1995). Simultaneous treatment with AZT and 3TC can suppress virus load about 10-fold in patients treated for up to one year (Eron *et al.* 1995; Staszewski 1995).

Several mathematical models have been developed to describe the population dynamics of virus replication following drug treatment and the emergence of resistant mutant. McLean *et al.* (1991) proposed that the short-term effect of AZT treatment is due to the predator–prey like interaction between virus and host cells, and that the CD4 cell increase following drug treatment is responsible for the resurgence of virus even in the absence of resistant mutants. McLean and Nowak (1992*a*) postulate that the rise of drug-resistant virus is caused by the increase in available target cells for HIV infection.

Nowak *et al.* (1991) study the effect of drug treatment on delaying progression to disease. Frost and McLean (1994), McLean and Frost (1995) and De Jong *et al.* (1996) describe models for the sequential emergence of resistant variants in AZT therapy. Wein *et al.* (1997) use a control theoretic approach for multi-drug therapies. Kirschner and Webb (1996) study the emergence of drug resistant virus in a model with CD4 and CD8 cells. De Boer and Boucher (1996) propose that reducing CD4 cell numbers can potentially prevent the emergence of drug-resistant virus. Analytic approximations for the rise of drug-resistant virus were derived by Nowak *et al.* (1997*a*,*b*).

12

SIMPLE ANTIGENIC VARIATION

Soon after the discovery of HIV, in the mid-1980s, it was realized that the virus was characterized by extensive genetic variability. Beatrice Hahn and George Shaw, at first working with Robert Gallo at the NIH and later at the University of Alabama at Birmingham, were among the first to show that isolates of HIV differ extensively both when taken from different patients and when taken from the same patient at different time points. It became clear that HIV was changing continuously during the course of an infection, that the virus entering a patient was not the same as the virus one or several years after infection.

What is the source of HIV's extensive genetic variation? Several factors contribute to the relentless change of HIV during individual infections: (i) the low replication accuracy of reverse transcription; (ii) recombination during reverse transcription; (iii) many rounds of replication in an individual host with many virus particles produced from one infected cell; (iv) selective pressure for variation as exerted by the immune response; (v) a range of different target cells which can be infected by HIV, but require specific adaptations of the virus.

HIV is not unique with respect to its enormous genetic variability. Other lentiviruses, such as simian immunodeficiency virus (SIV), visna medi virus, equine infectious anaemia virus, caprine arthritis encephalitis virus and feline immunodeficiency virus also display extensive genetic variation during single infections. A similar picture is currently emerging for hepatitis C virus infection.

Genetic variation is not uniform throughout the HIV genome. There is more variability in the *env* gene than in the *gag* and *pol* genes. Within the *env* gene there are five hypervariable regions. Much attention has been paid to the third hypervariable region, the V3 loop, because it has immunodominant properties. This region is about 30 amino acids long and contains epitopes for neutralizing antibodies, CD4 cell and CD8 cell responses. It also seems to play an important role in virus entry into the host cell and thereby influences cell tropism. A single amino acid substitution can restrict recognition by neutralizing antibodies. Such a point mutation may be within the V3 loop or somewhere else in the envelope protein (gp-120); in the second case a conformational change may prevent antibody binding. Thus genetic change in the envelope protein may help the virus to escape from antibody responses. Genetic variation which leads to changes in protein comformations seen by immune responses is called 'antigenic variation'.

Similarly, single-point mutations in the *env*, *gag*, *pol* or other genes of HIV can prevent recognition by T-cell responses. In 1991, Andrew McMichael, Rodney Phillips and colleagues at the University of Oxford provided the first evidence that genetic variation in the *gag* gene can lead to virus mutants that escape from cytotoxic T-cell responses.

In 1990, we proposed a theory according to which disease progression in HIV infection is an evolutionary process. We assumed that the asymptomatic phase is a highly dynamic process where the virus population and the immune system are in a steady-state. Given the dynamic nature of the virus–immune system interaction, where elementary processes happen on a very fast time-scale, the question was why it took many years for HIV to defeat the immune system. In other words, which process can slowly shift the balance between HIV and the immune system in favour of the virus? Our general idea was that the rapid genetic variation of the virus generates over time viral populations (quasispecies) which are more and more adapted to grow well in the micro-environment of a given patient. Selection will continuously favour viral mutants that have overall faster growth rates than their competitors. Faster growth rates may result from changes in the envelope protein that allow a more rapid entry of the virus particle into the target cell. They may also be a consequence of mutations in viral control genes that enhance the pace and quantity of viral particles produced from an infected cell. More importantly, however, for the evolutionary process we have in mind may be changes in viral proteins that avoid elimination of the virus by immune responses.

Thus the specific idea of our early HIV dynamics model was that virus evolution during individual infections continuously enables the virus to escape from immune responses. This process is known as antigenic variation, and we suggest that it generates the long-term dynamics that give rise to the overall pattern of disease progression in HIV infection.

The theory is based on two fundamental assumptions: (i) HIV is controlled by immune responses; by this we mean that the immune responses against HIV are capable of down-regulating virus abundance at least during the early stages of infection; (ii) HIV can mutate during the course of an infection, and some of these mutants can escape from certain immune responses.

With these two assumptions we formulate the basic model of antigenic variation in Section 12.1, which leads to a dynamic steady-state of virus load with a rapid and continuous turnover of virus. Virus load is held in balance by ongoing viral reproduction and elimination by immune responses. The level of virus load is a function of how fast the virus grows and how efficient the virus is eliminated by the immune system. Virus load also depends on antigenic diversity. Higher levels of antigenic diversity lead to a higher virus load. If HIV continuously generates new antigenic material, as Darwinian evolution and quasispecies theory would suggest, then virus load should increase as a consequence of virus evolution.

Of course, antigenic variation alone does not make an HIV model. Other pathogens display antigenic variation, but do not cause a disease like HIV infection. Therefore the fundamental third assumption of our model is: (iii) HIV can impair immune responses. The virus infects and kills CD4 positive T helper cells which have a central role in organizing immune responses to infectious agents. Thereby the virus weakens the immune response which is meant to overcome it. The most interesting consequence of adding this assumption to our model, as done in Section 12.2, is the emergence of a diversity threshold phenomenon. The immune system and the virus population are in a defined steady-state only if the antigenic diversity of the virus population is below a certain threshold value. If the antigenic diversity exceeds this threshold than the virus population can no longer be controlled by the immune system. This diversity threshold is not caused by any saturation in the capacity of the immune response as such, but rather is an intrinsic—and unusual—property of the dynamics of this peculiar 'host–parasite' system.

As we shall see, the model has three parameter regions. We distinguish between immune responses that are active against individual virus mutants (or strains), termed strain-specific responses, and mutants which can recognize all virus mutants in a patient, termed cross-reactive immune responses.

In the first parameter region, a single virus mutant can outgrow the effect of specific and cross-reactive immune responses. In this case, the immune system cannot control the virus population. There is immediately high virus load and rapid progression to disease and death. Note that this fast progression can occur in the absence of antigenic diversity.

In the second parameter region, the cross-reactive immune response is sufficiently strong to hold down the virus population without the help of strain-specific responses. In this case there is no diversity threshold. The virus is held in steady-state even for very high levels of antigenic diversity. Increasing antigenic diversity can increase virus load, but does not enable the virus to escape from the immune response altogether.

The third parameter region is characterized by a diversity threshold. Here cross-reactive responses are not sufficiently strong to down-regulate the virus population by themselves, but a combination of cross-reactive and strain-specific responses can hold in check every individual mutant. It is the accumulation of mutants which makes the virus escape eventually. As we shall see, increasing antigenic diversity essentially negates the effect of strain-specific immune responses.

12.1   The basic model of antigenic variation

The simplest model of antigenic variation describes a replicating viral (or other) pathogen which is opposed by strain-specific immune responses. Let v_i denote the population size of virus strain (or mutant) i, and let x_i denote the magnitude of the specific immune response against strain i. (In the previous chapters, x denoted uninfected cells. Here, and in the following chapters, x_i denotes the specific immune response against virus strain i.) Consider the following system of ordinary differential equations:

$$\frac{dv_i}{dt} = rv_i - px_iv_i, \quad i = 1, \ldots, n,$$

$$\frac{dx_i}{dt} = cv_i - bx_i, \quad i = 1, \ldots, n. \tag{12.1}$$

In the absence of immune responses, viral growth is exponential at rate, r. Immune responses are stimulated at the rate cv_i, which is proportional to the abundance of virus. Immune responses eliminate virus at the rate $px_i v_i$. Finally, immune responses decay in the absence of further stimulation at rate bx_i. In the model, there are n virus strains, which are opposed by n specific immune response. Each immune response, x_i, can only recognize virus strain v_i. Note that in this simple model we do not distinguish between infected cells and free virus. Furthermore, virus growth is not limited by target cell abundance. These simplifications allow an elegant mathematical treatment; dropping these simplifications does not change the essential results. (See Regoes *et al.* 1998.)

Figure 12.1 shows a computer simulation of the above model. We start with a single virus strain, v_1. Initially virus growth is exponential at rate r, but the virus stimulates the specific immune response, x_1, which reduces virus growth and eventually brings an end to the viral expansion. Virus load reaches a maximum value and subsequently starts to decline. Similarly, the immune response, x_1, reaches a maximum value and then declines. The system settles in damped oscillations to the equilibrium

$$v_1^* = \frac{br}{cp}, \quad x_1^* = \frac{r}{p}. \tag{12.2}$$

We assume, however, that mutation continously generates new viral strains which can escape from the specific immune responses. In the computer simulation of Fig. 12.1, the generation of new mutants is a stochastic process; the probability that a new mutant emerges in the time interval $[t, t + dt]$ is given by $P\, dt$, where P is the mutation rate. The simplest assumption is that P is a constant. It might be more realistic, however, to assume that P is proportional to the total virus load because the number of mutation events is proportional to the number of replication events.

In Fig. 12.1, the new variant, v_2, escapes from the immune response x_1 and grows unchecked initially, but it induces an immune response, x_2, which brings it down after some time. In the meanwhile, another escape mutant, x_3, has emerged and so on. The result is a sequence of antigenically different variants that all grow for some time before being controlled by immune responses. If there are n viral variants present in the system and if all are at the equilibrium value given by (12.2) then the total virus abundance, $v = \sum_{i=1}^{n} v_i$, is

$$v = \frac{brn}{cp}. \tag{12.3}$$

Thus, on average (smoothing out transient dynamics), virus load, v, is an increasing function of antigenic diversity, n.

Equation (12.1) defines the most basic model of antigenic variation. Each virus strain, v_i, is only controlled by one specific immune response, x_i. This means that the dynamics of any one strain are independent of all other strains, which is unrealistic. (Note that the differential equations describing the dynamics of one strain and its specific immune response are decoupled from the equations for other viral strains.)

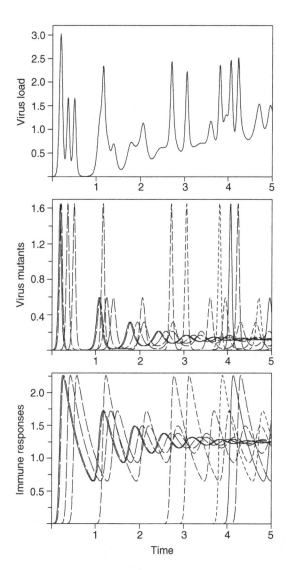

Fig. 12.1 Dynamics of the basic model of antigenic variation as defined by eqn (12.1). Each viral strain is only opposed by a strain-specific immune response. There is no cross-reactive immunity. Therefore the dynamics of each strain are independent of all other strains. Each strain rises to high abundance and is subsequently down-regulated in damped oscillations by a specific immune response. Virus load increases in an oscillatory fashion as new strains are being generated. The equilibrium virus load is an increasing function of antigenic diversity. The figure shows the total virus load, v, the abundance of individual strains, v_i, and the strength of specific immune responses, x_i. Parameter values are: $r = 2.5$, $p = 2$, $c = b = 0.1$. The infection starts with a single strain. The probability that a new mutant arises in the time interval $[t, t + dt]$ is given by Pdt with $P = 0.1$.

The most obvious extension of the basic model is to include a cross-reactive immune response which can recognise several (or all) virus mutants. Thus not all immune responses are strain-specific, and new antigenic variants do not completely escape from all existing immune responses in the patient. Denote by z the strength of a cross-reactive immune response which is active against all virus mutants in the system. This leads to the equations

$$\frac{dv_i}{dt} = v_i(r - px_i - qz), \quad i = 1, \ldots, n,$$

$$\frac{dx_i}{dt} = cv_i - bx_i, \quad i = 1, \ldots, n, \qquad (12.4)$$

$$\frac{dz}{dt} = kv - bz.$$

The cross-reactive immune response, z, is stimulated by each virus mutant at rate kv_j and decays at the rate bz. The main consequence of this model extension is that new antigenic variants are not completely escaping from all existing immune responses but are partly recognized, and therefore do not rise to the same abundance as the original virus mutants. The dynamics of individual strains are no longer independent from each other. A computer simulation is shown in Fig. 12.2.

For n virus mutants, the equilibrium total virus load is given by

$$v^* = \frac{brn}{cp + kqn}. \qquad (12.5)$$

Once again, average viral load is an increasing function of antigenic diversity, n, but saturates at high values of n. The maximum possible equilibrium virus load is $v^*_{max} = (br)/(kq)$, which represents the equilibrium virus load in the presence of the cross-reactive immune response alone. Thus increasing antigenic diversity eliminates the effect of strain-specific immunity.

In the model given by eqn (12.4), the individual virus strains are defined as being different with respect to strain-specific immune responses, but all strains are recognized by the cross-reactive immune response. In reality the situation is considerably less clear-cut, because viruses usually have several different epitopes which can be recognized either by antibodies or T cells. Some virus mutants may differ in one epitope, but coincide in others. This means that a given antibody or cytotoxic T lymphocyte may be able to recognize a number of different virus strains, but fail to recognize others. There is a variety of more or less cross-reactive and strain-specific immune responses. The model only considers the two extreme possibilities of completely cross-reactive and completely strain-specific immune responses. The whole spectrum of immune responses is covered by assigning parameter values to balance the relative importance of strain-specific versus cross-reactive responses.

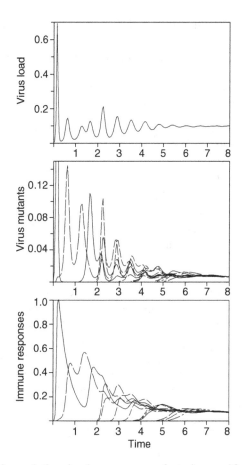

Fig. 12.2 Antigenic variation in the presence of strain-specific and cross-reactive immune responses. The simulation begins with a single viral strain, which induces both strain-specific and cross-reactive immune responses. Subsequent strains escape from the strain-specific response, but not from the cross-reactive response. The equilibrium virus load is an increasing function of antigenic diversity, but saturates at high levels of antigenic diversity. The simulation is based on eqn (12.4) with the parameter values $r = 2.5$, $p = 2$, $q = 2.4$, $c = k = 1$, $b = 0.1$. The probability that a new mutant arises in the time interval $[t, t + dt]$ is given by $P dt$ with $P = 0.1$.

12.2 Antigenic variation of HIV: diversity threshold

In the previous section, we analysed general models of antigenic variation, which were not specifically designed for HIV infection, but describe in principle the dynamics of any infectious agent displaying antigenic variation. In this section, we will include a term that is a specific property of HIV infection: the virus can infect and eliminate CD4 positive

T helper cells, which constitute an essential arm of the immune system. Therefore, HIV can impair immune responses.

The simplest mathematical model that captures this specific property of HIV is the following:

$$\frac{dv_i}{dt} = v_i(r - px_i - qz), \quad i = 1, \dots, n,$$

$$\frac{dx_i}{dt} = cv_i - bx_i - uvx_i, \quad i = 1, \dots, n, \tag{12.6}$$

$$\frac{dz}{dt} = kv - bz - uvz.$$

As before, v_i denotes the population size of virus mutant i; x_i denotes the immune response specifically directed against virus strain i; and z denotes the cross-reactive immune response directed against all different virus strains. The total number of different virus strains is given by n; mutational events occur throughout the infection and therefore increase this number, n, as time goes by. As before we use the notation $v = \sum v_i$ where v denotes the total population of virus.

The parameter r denotes the average rate of replication of all different virus strains; p specifies the efficacy of the strain-specific immune responses and c specifies the rate at which they are evoked; similarly q specifies the efficacy of group-specific immune responses and k the rate at which they are evoked. In the absence of further stimulation, the immune response decays at a rate given by the constant b. HIV and other lentiviruses can impair immune responses, by killing CD4 positive T helper cells, which help B cells and CD8 positive cytotoxic T cells to mount immune responses against the virus. We summarize this effect of HIV in the loss terms, $-uvx_i$ and $-uvz$. Thus the parameter u characterizes the ability of the virus to impair immune responses; by depleting CD4 cells the virus impairs indirectly B-cell and cytotoxic T-cell mediated immune responses.

From eqns (12.6) we obtain an equation for the rate at which the total virus population changes:

$$\frac{dv}{dt} = \frac{v}{b + uv}[rb - v(kq + cpD - ru)]. \tag{12.7}$$

For this we have assumed that the individual x_i converge to their steady-state levels $x_i^* = cv_i/(b + uv)$, and that z converges to $z^* = kv/(b + uv)$, on time scales short compared with those on which the total virus population changes. D denotes the Simpson index, $D = \sum (v_i/v)^2$, which is an inverse measure for viral diversity: if there is only one virus strain present then $D = 1$; if there are n strains present, all of them exactly at the same abundance, then $D = 1/n$. D is always between 0 and 1; it is actually the probability that two viruses chosen at random belong to the same strain. The concept of a virus strain, v_i, is well defined in the mathematical model. It is simply a subpopulation of viruses that are all recognized by the same strain-specific immune response, x_i.

From eqn (12.7) we see that v converges towards the steady-state

$$v^* = \frac{rb}{kq + cpD - ru}. \tag{12.8}$$

The product kq specifies the efficacy of the cross-reactive immune responses, such as antibodies or CTLs directed at epitopes that are conserved between different virus strains. The product cpD denotes strain-specific immune responses, such as antibodies or CTLs directed at variable regions. The efficacy of these strain-specific responses depends on the antigenic diversity of the virus population. Equation (12.8) shows that increasing diversity (decreasing D) increases the total population size of the virus.

The model has three distinct parameter regions, which correspond to three qualitatively different courses of infection.

1. *There is no asymptomatic phase: the virus population immediately replicates to high levels.* This happens if

$$ru > kq + cp. \qquad (12.9)$$

In this case, viral replication, r, and/or cytopathic effects, u, are large compared to the combined effects of cross-reactive and strain-specific immune responses, $kq + cp$. Hence $dv/dt > 0$ for all v, and the immune response is unable to control the virus, which replicates to high levels and may induce acute disease and death within a short time. No antigenic variation may be observed in this case, because of selection for the fastest growing virus strain. The immune system does not have time to select for antigenic diversification.

2. *Chronic infection, but no disease.* This happens if the cross-reactive immune response alone is sufficiently large to control the virus population. In mathematical terms this means

$$kq > ru. \qquad (12.10)$$

That is, the cross-reactive immunity, kq, is large compared to viral replication and killing of immune cells, ru. Here the equilibrium of eqn (12.8) will be attained, no matter how small D becomes, and the immune response is able to suppress viral concentrations to very low levels. The time average of the virus population size during the chronic infection is given by eqn (12.8) and depends on: (i) the rate of virus replication; (ii) the extent of impairment of immune functions; and (iii) the efficacy of the immune responses. The effect of the strain-specific immune responses depends on the overall antigenic diversity. If there is a large amount of variation, then there are always some escape mutants that can replicate in the presence of an immune response, and there is an increase in the total virus population size. Thus, the virus level increases with increasing diversity, but in this parameter regime there is no critical diversity threshold beyond which eqn (12.8) ceases to give sensible results. The immune system is in principle able to control the virus population indefinitely (virus levels increase with increasing diversity, but in a controlled manner). Figure 12.3 illustrates the dynamics of the infection for this parameter region.

3. *Chronic infection and disease after a (possibly long) incubation period.* The third, and most interesting, situation arises when the combined effects of cross-reactive and

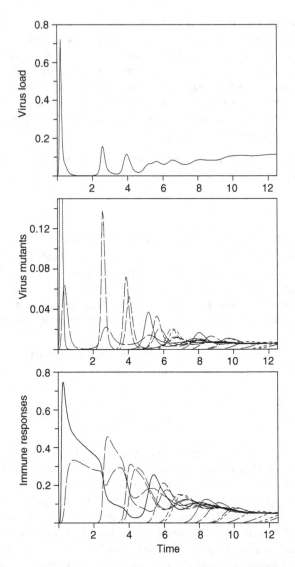

Fig. 12.3 A strong cross-reactive immune response (directed at the conserved epitopes of the virus) can lead to a chronic infection without development of disease. This situation occurs if the cross-reactive response alone is sufficiently strong to control the virus population, that is if $kq > ru$. The computer simulation is based on eqn (12.6) with the parameter values $r = 2.3$, $p = 2$, $q = 2.4$, $c = k = u = 1$, $b = 0.01$. The infection starts with a single strain. The probability that a new mutant arises in the time interval $[t, t + dt]$ is given by Pdt with $P = 0.1$.

strain-specific immune responses are able to control the virus replication of any one individual strain, but the cross-reactive response alone is unable to do so (in contrast with case 2). In mathematical terms this means

$$kq + cp > ru > kq. \tag{12.11}$$

Figure 12.4 shows the dynamics of the model for this parameter region. Now the strain-specific immune responses play an important role. If the antigenic diversity is low (D large) then the total population size is regulated to some equilibrium value (given by eqn (12.8)). If antigenic diversity is high (D low) then the denominator in eqn (12.8) becomes very small, and hence the virus population size very large. The critical transition occurs when

$$D < \frac{ru - kq}{cp}. \tag{12.12}$$

Alternatively, in this simplest case where all strains have the same parameter values, we have $D = 1/n$, and the threshold number of viral strains is given by

$$n > \frac{cp}{ru - ka}. \tag{12.13}$$

Beyond this point, the total virus population grows unboundedly. Equation (12.12) gives the diversity threshold. Once this threshold of viral diversity is exceeded, then the virus population escapes from control by the immune response and tends to arbitrarily high densities. This process may be interpreted as the development of immunodeficiency disease, which is characterized by high virus counts and depletion of CD4+ cells. During the asymptomatic phase, on the other hand, the diversity is increasing, but the immune system is able to control viral densities and to maintain CD4 cell levels.

12.3 Three kinds of observed lentivirus infections

The closest relative of HIV is SIV. Multiple isolates of SIVs have been obtained from a variety of non-human primates. These include African green monkeys (AGM), pig tailed macaques, cynomolgus monkeys and sooty mangabeys. There are large variations in the pathogenesis of different SIV isolates from the same species of monkey, or in the same isolate from different monkey species.

It is tempting to compare the three qualitatively different solutions of our mathematical model with the three different patterns of lentivirus infections.

1. Rapid progression to disease and death essentially without prolonged asymptomatic period has been observed in some HIV infected patients. In terms of our model these rapid progressors fall into parameter region 1: virus replication in these patients outruns the combined effect of strain-specific and cross-reactive immune responses; the immune system is unable to control the fast replicating virus. There is immediate development of

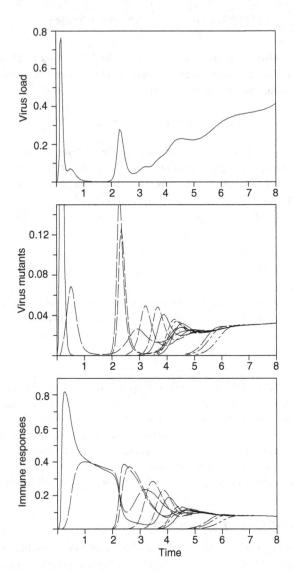

Fig. 12.4 A diversity threshold occurs if the cross-reactive immune response by itself is unable to control the virus population, but a combination between cross-reactive and strain-specific responses can control any one strain. In mathematical terms, this means $kq + cp > ru > kq$. Increasing antigenic diversity enables the virus population to escape from the immune response after a long incubation period. The computer simulation is based on eqn (12.6) with the parameter values $r = 2.5, p = 2, q = 2.4, c = k = u = 1, b = 0.01$; with these parameters, the diversity threshold is reached when the number of viral strains exceeds $n = 20$. The infection starts with a single strain. The probability that a new mutant arises in the time interval $[t, t + dt]$ is given by $P dt$ with $P = 0.1$.

high virus load and progression to disease. Antigenic variation is not necessary in this case. There are also acutely lethal variants of SIV that kill certain animals within a few weeks after infection.

2. Persistent infection without development of immunodeficiency disease appears to be the case for all lentivirus infections in their natural hosts. For example, about 50% of wild AGMs seem to harbour SIV, but disease has not been noticed. Similar observations have been made for other SIV or FIV infections in their natural hosts. The apathogenicity of these infections is unexplained. In terms of our model, these infections fall into parameter region 2: the cross-reactive immune responses alone are capable of controlling the virus; antigenic variation does not lead to complete escape from immune control. Thus the virus may show high levels of genetic and antigenic variation, but still be maintained at low levels.

3. The third type of infection—with its long asymptomatic phase followed by lethal immunodeficiency—is observed for HIV infection in humans and experimental SIV infections, where the virus from one monkey species is used to infect another species. In correspondence to parameter region 3 of our model, we suggest that in this case the cross-reactive immune response alone is not capable of down-regulating the virus, but the combined action of strain-specific and cross-reactive responses is required. Antigenic variation eliminates the effect of the strain-specific responses and ultimately leads to very high virus load and development of disease.

Which outcome is the case for a particular lentivirus infection depends on both the host and the virus.

Viral properties that favour pathogenicity (parameter regions 1 and 3) are the overall replication rate of the virus in a given host and its ability to impair immune responses (kill CD4 cells), corresponding to our parameters r and u, respectively. Both factors may crucially depend on the particular cell tropism of the virus. If the virus grows preferentially in macrophages than CD4 T cells, then this may reduce both r and u.

Host properties that work against pathogenicity in our model (parameter region 2) are the efficacy of the immune response against the virus, defined by the response parameters k and c, and the efficacy parameters q and p. In apathogenic lentivirus infections, hosts may have been selected for strong cross-reactive immune responses against the virus. In addition, host genetics may influence the overall viral growth rate and immune pathogenicity. Genetic variation in receptor genes can bias the virus toward a cell tropism which only supports low levels of virus and/or low levels of immune impairment. Note also that low values of b, which is in some sense an immune memory parameter, leads to low virus load. (See Wodarz et al. (2000) for a specific theory.)

We think there is potentially much to be learned from setting up a detailed study for a systematic comparison of pathogenic and apathogenic lentivirus infections, in order to determine if major differences exist in virus load, viral turnover rates, cell-tropism or immune responses. Such a comparative analysis should ultimately be the decisive approach toward a detailed understanding of HIV pathogenesis.

12.4 Further reading

Nowak *et al.* (1990, 1991) describe models for the population dynamics of virus infections and propose an evolutionary mechanism for HIV disease progression. See also Nowak (1992*a*) and Nowak and May (1991, 1992, 1993), Nowak and McLean (1991). Early publications on the genetic variation of HIV-1 during single infections are Hahn *et al.* (1986), Fisher *et al.* (1988), Saag *et al.* (1988), Meyerhans *et al.* (1989), Balfe *et al.* (1990), Holmes *et al.* (1992) and Weiss *et al.* (1986). For genetic variation of HIV-2 see Schulz *et al.* (1990); of SIV see Burns and Desrosiers (1991), Overbaugh *et al.* (1991) and Baier *et al.* (1980); of visna virus see Clements *et al.* (1980); of EIAV see Salinovich *et al.* (1986); and of CAEV see Ellis *et al.* (1987). For genetic and antigenic variation of the V3 loop we refer to Simmonds *et al.* (1990), Goudsmit (1988), Moore and Nara (1991), Wolfs *et al.* (1991) and Nara *et al.* (1990). Escape from T-cell recognition is described by Phillips *et al.* (1991), McAdam *et al.* (1995), and McMichael *et al.* (1996). For analysing selection pressure in the V3 loop we refer to Bonhoeffer *et al.* (1995). Price *et al.* (1997) study a patient with an immunodominant response against Nef during primary HIV-1 infection. Within a few months of infection the wild-type virus was replaced by mutants that escape recognition by the patient's CTL directed at this epitope. Additional studies of antigenic variation in CTL epitopes in HIV-1 infection are Borrow *et al.* (1997*a,b*) and Goulder *et al.* (1997). Selection of CTL escape variants *in vivo* was first demonstrated by Pircher *et al.* (1990) in lymphocytic choriomeningitis virus (LCMV) infection. See also Moskophides and Zinkernagel (1995). For variation of HTLV-1 see Niewiesk *et al.* (1994, 1995, 1996). McKnight and Chorphar (1995) describe immune escape and tropism of HIV.

13

ADVANCED ANTIGENIC VARIATION

In this chapter, we discuss more complicated models of antigenic variation. In the simple models of Chapter 12, all viral strains had the same kinetic parameters. Now we assume that strains may differ in all their characteristic parameters, such as replication rate, immunological properties and cytopathicity. This leads to important new insights.

If viral strains differ in their replication rates, then there is an element of competition. In the absence of strain-specific immune responses, the fastest replicating strain will win and all others go to extinction. Therefore antigenic diversity in a patient is not only a consequence of viral mutation, but also of the patient's immune system selecting for antigenic variation. More specifically, we will show that cross-reactive immune responses provide a selection pressure *against* antigenic variation, while strain-specific responses select *for* antigenic variation. The level of antigenic diversity in a patient reaches a steady-state between competition among strains (that is, selection against diversity) and selection for diversity. We will show that the amount of antigenic diversity in a patient is determined by the ratio of cross-reactive versus strain-specific immune responses. Enhancing cross-reactive immunity should reduce antigenic diversity. Therefore the immune system can provide selection pressure for and against antigenic diversification.

Although the models suggest that sequentially, in all individual patients, antigenic diversity increases virus load, the correlation between antigenic diversity and virus load in cross-sectional comparisons among different patients can be complicated. If patients differ mainly in their strain-specific immune responses, then we expect a *negative* correlation between antigenic diversity and virus load. The more virus in a patient the less diversity. This is intuitively obvious: a patient with weak strain-specific immune responses allows a high virus load but provides little selection pressure for antigenic variation. A patient with strong strain specific responses controls the virus population to low levels, but selects for higher diversity. Conversely, if patients differ mostly in their cross-reactive responses against the virus, then the correlation between virus load and antigenic diversity in a cross-sectional comparison among patients should be positive: a patient with a weak cross-reactive response allows a high virus load with weak selection pressure against diversity.

Experimental findings have provided some evidence for low *genetic* variation in HIV-infected patients with high virus load and rapid progression to disease. While it is problematic to link *genetic* variation directly to *antigenic* variation, such findings are in clear agreement with the diversity threshold idea: these rapid progressors have weak strain-specific immune responses and therefore do not select for high levels of antigenic diversity.

As in Chapter 12, we will first analyse a model without immune function impairment (Section 13.1). Such a model does not have a diversity threshold, but antigenic diversity increases virus load during individual infections. This model applies to all viruses (or more generally parasites) that cause persistent infections and can generate antigenic variation. In Section 13.2, we include immune function impairment. Section 13.3 summarizes the results of the advanced models of antigenic variation.

13.1 Immune response can select for or against antigenic diversity

In the Section 12.1, we considered the simplest models of antigenic variation. In particular, all virus mutants and immune responses had the same rate constants. Now we analyse the same model, but assume that different virus mutants have different replication rates and are opposed by immune responses of different strength. Consider the following system of ordinary differential equations:

$$\frac{dv_i}{dt} = v_i(r_i - p_i x_i - q_i z), \quad i = 1, \ldots, n,$$

$$\frac{dx_i}{dt} = c_i v_i - b_i x_i, \quad i = 1, \ldots, n, \tag{13.1}$$

$$\frac{dz}{dt} = \sum_{j=1}^{n} k_j v_j - bz.$$

As before, v_i denotes the abundance of virus strain (or mutant) i, x_i the magnitude of the specific immune response against strain i, and z the magnitude of the cross-reactive immune response directed at all strains. There are n strains; each strain is characterized by a set of parameters: r_i denotes the replication rate of strain i; p_i the rate at which the strain is killed by specific immune responses; q_i the rate at which the strain is killed by cross-reactive immune responses; c_i and k_i are the rates at which strain i induces specific and cross-reactive immune responses. Specific immune responses, x_i, decline at rate b_i, the cross-reactive immune response declines at rate b.

The system has 2^n possible equilibria given by the equations:

$$v_i = 0 \quad \text{or} \quad r_i = \frac{c_i p_i}{b_i} v_i + \frac{q_i}{b} \sum_{j=1}^{n} k_j v_j, \tag{13.2}$$

$$x_i = \frac{c_i}{b_i} v_i, \tag{13.3}$$

$$z = \frac{1}{b} \sum_{j=1}^{n} k_j v_j. \tag{13.4}$$

It is convenient to rescale the virus concentrations $v_i' = k_i v_i$, which leads to

$$r_i = \frac{c_i p_i}{b_i k_i} v_i' + \frac{q_i}{b} v'. \tag{13.2a}$$

Here $v' = \sum_{j=1}^{n} v'_j$. We will now show that among the 2^n possible equilibria, there is only one stable equilibrium. For the stable equilibrium we must have that all strains with $v'_i = 0$ must be unable to invade, i.e. their transversal eigenvalue $\partial \dot{v}'_i / \partial v'_i = r_i - p_i x_i - q_i z$ at this equilibrium must be negative. (The transversal eigenvalue denotes the per capita growth rate of virus strain i at the relevant boundary equilibrium where $v_i = 0$. 'Transversal eigenvalue' is an expression coined by Josef Hofbauer and Karl Sigmund.) We assume that there is no specific immune response in the beginning and therefore $x_i = 0$. This leads to the condition $r_i - q_i z < 0$ or $r_i/q_i < v'/b$. It follows that we can rank the virus strains according to their ratio r_i/q_i. Without loss of generality we label the strains such that

$$r_1/q_1 > r_2/q_2 > \cdots > r_n/q_n. \tag{13.5}$$

This rank order implies that if a certain equilibrium cannot be invaded by strain i, it can also not be invaded by any strain with an index greater than i. Thus it is sufficient to consider equilibria of the form: $v'_i > 0$ for $i = 1, \ldots, m$ and $v'_i = 0$ for $i = m+1, \ldots, n$. Let us call such an equilibrium E_m and let us denote the total (rescaled) virus population size at this equilibrium by V_m. By summation of (13.2) over $i = 1, \ldots, m$ we obtain

$$V_m = \sum_{i=1}^{m} k_i \alpha_i \Bigg/ \left(1 + \sum_{i=1}^{m} k_i \beta_i\right) \tag{13.6}$$

with $\alpha_i = b_i r_i / (c_i p_i)$ and $\beta_i = b_i q_i / (b c_i p_i)$.

Equilibrium E_m is only stable if two conditions are fulfilled. First, strain m must be able to invade equilibrium E_{m-1}. This is the case if

$$\frac{r_m}{q_m} > \frac{1}{b} V_{m-1}, \tag{13.7}$$

which is equivalent to $V_m > V_{m-1}$. Second, strain $m + 1$ must be unable to invade equilibrium E_m. The condition that strain $m + 1$ cannot invade is

$$\frac{r_{m+1}}{q_{m+1}} < \frac{1}{b} V_m. \tag{13.8}$$

It is straightforward to show that this condition is equivalent to V_m being larger than V_{m+1}.

Therefore the unique stable equilibrium of system (13.1) is characterized by the highest rescaled virus load among all equilibria. If all strains stimulate the cross-reactive response at the same rate, $k_i = k$, then the unique stable equilibrium is characterized by the highest viral load among all possible equilibria. In this sense, selection maximizes viral load. If there are differences in the stimulation of the cross-reactive response, then there is no longer selection for maximum virus load, and a strain which induces a lower viral load can outcompete a strain which induces a higher viral load.

Thus the way to define the stable equilibrium is the following: order all strains according to their ratios r_i/q_i. Then determine the index m which maximizes V_m, the rescaled virus population size. If we go back to our original variables v_i, we can write down the equilibrium load, v, by using (13.2) and (13.6). With m defined in this way, the stable equilibrium is given by

$$v_i = \alpha_i - \beta_i V_m, \tag{13.9}$$

and the total viral load by

$$v = \sum_{i=1}^{m} v_i = \sum_{i=1}^{m} \alpha_i - \sum_{i=1}^{m} \beta_i V_m. \tag{13.10}$$

We have now characterized the dynamical behaviour of system (13.1) and calculated the unique globally stable equilibrium. In order to gain further analytical insight, particularly into the relation between virus load and diversity, we consider the special case where $p_i = p, q_i = q, b_i = b$, and $k_i = k$ for all strains. Thus we analyse the simplified system

$$\frac{dv_i}{dt} = v_i(r_i - px_i - qz), \quad i = 1, \ldots, n,$$

$$\frac{dx_i}{dt} = c_i v_i - bx_i, \quad i = 1, \ldots, n, \tag{13.11}$$

$$\frac{dz}{dt} = kv - bz.$$

We rank the strains according to $r_1 > r_2 > \cdots > r_n$. At equilibrium the individual virus strains have the abundances

$$v_i = \frac{br_i - kqv}{c_i p}, \quad i = 1, \ldots, m, \tag{13.12}$$

and $v_i = 0$ for $i = m + 1, \ldots, n$. The total virus population size is

$$v = b \sum_{i=1}^{m} \frac{r_i}{c_i} \bigg/ \left(p + kq \sum_{i=1}^{m} \frac{1}{c_i} \right), \tag{13.13}$$

and m is the largest integer fulfilling

$$r_m > kq \sum_{i=1}^{m-1} \frac{r_i}{c_i} \bigg/ \left(p + kq \sum_{i=1}^{m-1} \frac{1}{c_i} \right). \tag{13.14}$$

Looking at eqn (13.12), it is clear that $br_i - kqv$ is a declining sequence with i and that $br_i - kqv > 0$ if $i \leq m$ and $br_i - kqv < 0$ if $i > m$. For a large number of strains, m, a good approximation is $br_m - kqv = 0$. Hence, for the relation between virus load, v, and diversity, m, we derive the result

$$v = \frac{b}{kq} r_m. \tag{13.15}$$

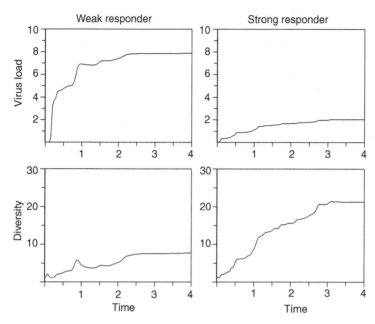

Fig. 13.1 Evolution of virus load and diversity in two patients. One patient has a weak strain-specific immune response, while the other patient has a strong strain-specific immune response. In the weak responder, virus load raises quickly to high levels, but the antigenic diversity remains low. In the strong immune responder, virus load is maintained at low levels while the immune response selects for high diversity. In both cases, antigenic diversity increases virus load. For the computer simulation, we use system (13.11) with r_i randomly chosen from a uniform distribution between 0 and 1, $k = 0.1$, $p = 1$, $q = 1$, $b = 1$, and $c_i = 0.1$ (for all strains i) for the weak responder and $c_i = 5$ for the strong responder. There is a constant probability over time to produce new antigenic variants, which increases the dimension n of the system. Virus diversity is given by the inverse of the Simpson index, D, which is defined as $D = \sum (v_i/v)^2$.

Figure 13.1 shows a computer simulation of the model described by eqn (13.11). New virus strains are added over time which increases virus diversity and virus load. The figure shows infection dynamics in two patients. One patient has weak strain-specific immune responses, while the other patient has strong strain-specific immune responses. In the weak responder virus load soon grows to high levels, but there is little selection pressure for antigenic variation. In the strong responder, virus load is down-regulated to low levels, but the immune system provides strong selection for variation.

The result can be understood in terms of competitive exclusion. In the absence of strain-specific immunity, only the strain with the highest replication rate would win. Strain-specific immunity down-regulates the abundance of the fastest replicating strains and therefore allows other strains to persist as well.

13.1.1 *Cross-sectional comparisons*

In a cross-sectional comparison among different patients, do we expect a positive or negative correlation between virus load and antigenic diversity? If individual patients differ in the strength of their specific immune response (parameters c_i and p) and in the intrinsic viral replication rates, r_i, but have the same cross-reactive immune response (parameters k and q), then eqns (13.14) and (13.15) predict a negative correlation between virus load and diversity. Patients with stronger strain-specific immune response (high c_i and p) select for more diversity (higher m, lower r_m) and lower virus load, v (Fig. 13.2). Equation (13.15) also gives an inverse correlation between load and diversity, if the only difference among patients are the replication rates, r_i, of individual virus strains.

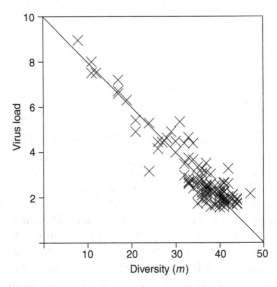

Fig. 13.2 A cross-sectional comparison among different patients gives an inverse correlation between viral load and diversity, if the patients differ in their strain-specific immune response. The figure shows equilibrium viral loads of system (13.11) as given by eqn (13.13). Individual virus strains have replication rates r_i which are taken from a uniform random distribution on the interval (0, 1). All patients have the same cross-reactive response, $k = 0.1$, but different strain-specific responses, c, ranging from 0 to 1. (We assume that $c_i = c$ for all strains.) The other parameters are: $p = q = b = 1$. There are $n = 50$ strains, and the figure shows 100 patient, each characterized by a ×. The left side of the figure shows virus load, v, versus virus diversity in terms of numbers of strains, m. The continuous line represents the approximation given by eqn (13.15), with a continuous approximation for $r_m = \rho(m) = 1 - m/n$. The right-hand side shows virus load versus the inverse of the Simpson index as a measure for diversity. The Simpson index is defined as $D = \sum (v_i/v)^2$.

If, on the other hand, k and q vary among patients while c_i and p are constant, then we must calculate the relation between kq and m before eqn (13.15) can provide an answer. We get an elegant analytical approximation if we assume $c_i = k$ for all strains. (This simplification essentially means that all immune responses are induced at the same rate.) Then the index m is defined as the largest integer which fulfills

$$r_m > \sum_{i=1}^{m-1} \frac{r_i}{\phi + m}, \tag{13.16}$$

where $\phi = cp/(kq)$ is essentially the ratio of strain-specific over cross-reactive immune responses. (We have also made the approximation that $m - 1 \approx m$.) If there is a very large number of strains, we can make a continuous approximation, $r_i = \rho(i)$, and the index m can be defined as the solution of the equation

$$\rho(m) = \frac{1}{\phi + m} \bar{\rho}(m) m, \tag{13.17}$$

where $\bar{\rho}(m)$ is the average replication rate for all strains with index up to m. If we consider for the r_i a uniform random distribution in the interval $(0, R)$, then we obtain $\rho(m) = R(1 - m/n)$ and $\bar{\rho}(m) = R[1 - m/(2n)]$. Equation (13.17) leads to

$$m^2 + 2\phi m - 2\phi n = 0. \tag{13.18}$$

For a given ϕ and n the number of strains at equilibrium is

$$m = -\phi + \sqrt{\phi^2 + 2n\phi}. \tag{13.19}$$

From eqn (13.18), we also get $kq = 2cp(n - m)/m^2$, which can be combined with eqn (13.15) to give

$$v = \frac{bR}{2cpn} m^2. \tag{13.20}$$

Hence, with the assumption of r_i following a uniform distribution we obtain a direct correlation between the virus load (v) and the square of virus diversity (m^2) if individual patients differ only in their cross-reactive responses against the virus (Fig. 13.3).

13.2 Antigenic variation of HIV with different parameter values

Next we analyse the model of Section 12.2, but allow for the possibility that different viral mutants have different replication rates and immunological properties.

$$\frac{dv_i}{dt} = v_i(r_i - p_i x_i - q_i z), \quad i = 1, \ldots, n,$$

$$\frac{dx_i}{dt} = c_i v_i - \left(b + \sum_{j=1}^{n} u_j v_j \right) x_i, \quad i = 1, \ldots, n, \tag{13.21}$$

$$\frac{dz}{dt} = \sum_{j=1}^{n} k_j v_j - \left(b + \sum_{j=1}^{n} u_j v_j \right) z.$$

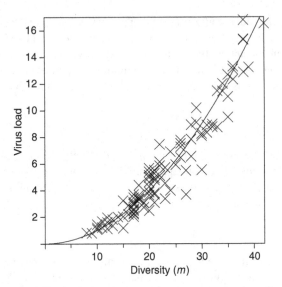

Fig. 13.3 If patients differ in their cross-reactive response against the virus, the model
predicts a positive correlation between virus load and diversity. Parameter values:
$p = q = b = 1$, $n = 50$, 100 patients, r_i from a uniform distribution on $(0, 1)$,
$c = 1$, and k from an exponential distribution with mean 0.2 (simply to get an equal
density of points over the interval). The continuous line indicates the analytical
approximation given by eqn (13.20).

In this model, x_i and z denote immune responses which require CD4 cell help. HIV
impairs such immune responses by reducing CD4 cell number and function. This is
described by the terms which contain $\sum u_j v_j$. Again without loss of generality, we rank
the strains such that $r_1/q_1 > r_2/q_2 > \cdots > r_n/q_n$. There are two possibilities. Either
the total virus population size, $v = \sum_i v_i$, grows to infinity or there is a unique stable
equilibrium given by

$$v_i = \frac{b}{c_i p_i \Lambda} \left[r_i \left(1 + \sum_{j=1}^{m} \frac{k_j q_j}{c_j p_j} \right) - q_i \sum_{j=1}^{m} \frac{k_j r_j}{c_j p_j} \right], \tag{13.22}$$

$$x_i = \frac{1}{p_i} \left[r_i - q_i \sum_{j=1}^{m} \frac{k_j r_j}{c_j p_j} \middle/ \left(1 + \sum_{j=1}^{m} \frac{k_j q_j}{c_j p_j} \right) \right], \tag{13.23}$$

$$z = \sum_{j=1}^{m} \frac{k_j r_j}{c_j p_j} \middle/ \left(1 + \sum_{j=1}^{m} \frac{k_j q_j}{c_j p_j} \right), \tag{13.24}$$

with

$$\Lambda = \left(1 + \sum_{j=1}^{m} \frac{k_j q_j}{c_j p_j} \right) \left(1 - \sum_{j=1}^{m} \frac{r_j u_j}{c_j p_j} \right) + \sum_{j=1}^{m} \frac{k_j r_j}{c_j p_j} \sum_{j=1}^{m} \frac{q_j u_j}{c_j p_j}. \tag{13.25}$$

The number m is given by the largest index which fulfills

$$\frac{r_m}{q_m} > \sum_{j=1}^{m-1} \frac{k_j r_j}{c_j p_j} \bigg/ \left(1 + \sum_{j=1}^{m-1} \frac{k_j q_j}{c_j p_j}\right). \tag{13.26}$$

For the total virus load at equilibrium, we obtain

$$v = \frac{b}{\Lambda}\left[\sum_{j=1}^{m} \frac{r_j}{c_j p_j}\left(1 + \sum_{j=1}^{m} \frac{k_j q_j}{c_j p_j}\right) - \sum_{j=1}^{m} \frac{q_j}{c_j p_j} \sum_{j=1}^{m} \frac{k_j r_j}{c_j p_j}\right]. \tag{13.27}$$

Whether this equilibrium exists or the virus population grows uncontrolled can be determined in the following way: check inequality (13.26) for increasing values of m. If the inequality holds for a particular value of m, then check if the corresponding virus load, v, given by eqn (13.27) is positive. If it is positive, augment m by 1 and check (13.26) again. The algorithm ends either if an m is found such that $v < 0$, or if an m is found such that inequality (13.26) is violated. In the first case the virus population grows to infinity; in the second case there is a stable equilibrium given by eqns (13.22)–(13.24). Note that v in eqn (13.27) can only be negative if Λ is negative, hence $\Lambda < 0$ is essentially a 'diversity threshold' condition.

As a special case of eqns (13.22)–(13.25), we get the complete equilibrium solution to the basic model of the previous section (eqn (13.1)) for $u_i = 0$ for all i. Interestingly this only affects Λ and v_i, but x_i, z, and v_i/v are not affected by immune function impairment. Note also that $\Lambda > 0$ if $u_i = 0$ for all i. Hence the diversity threshold is a feature unique to viruses that impair the immune response. (But for other viruses there is still a 'diversity advantage' in the sense that increasing antigenic diversity increases virus load.)

We will now study a simplified model to gain some analytical insight into the relation between virus load and diversity in HIV infections. As above, let us consider the special case where all strains have the same parameters p, q, k, b, and u and only differ in their replication rates, r_i, and their rate of stimulating the strain-specific response, c_i. This leads to

$$\frac{dv_i}{dt} = v_i(r_i - px_i - qz), \quad i = 1, \ldots, n,$$

$$\frac{dx_i}{dt} = c_i v_i - (b + uv)x_i, \quad i = 1, \ldots, n, \tag{13.28}$$

$$\frac{dz}{dt} = kv - (b + uv)z.$$

The individual viral strains have the equilibrium frequencies

$$v_i = \frac{1}{c_i p}[r_i(b + uv) - kqv], \quad i = 1, \ldots, m, \tag{13.29}$$

and $v_i = 0$ for $i = m + 1, \ldots, n$. The total virus load at equilibrium is given by

$$v = b \sum_{i=1}^{m} \frac{r_i}{c_i} \Bigg/ \left(p + kq \sum_{i=1}^{m} \frac{1}{c_i} - u \sum_{i=1}^{m} \frac{r_i}{c_i} \right). \tag{13.30}$$

The number of strains, m, is defined as the largest integer which fulfills

$$r_m > kq \sum_{i=1}^{m-1} \frac{r_i}{c_i} \Bigg/ \left(p + kq \sum_{i=1}^{m-1} \frac{1}{c_i} \right). \tag{13.31}$$

The requirement that the denominator of eqn (13.30) is positive gives the diversity threshold condition. If

$$p + kq \sum_{i=1}^{m} \frac{1}{c_i} > u \sum_{i=1}^{m} \frac{r_i}{c_i}, \tag{13.32}$$

then the virus population converges to the finite equilibrium given above. If the reverse holds then the immune responses cannot control the virus population. In mathematical terms, the virus population grows to infinity. The cross-reactive response converges to the value k/u, and each strain which fulfills $r_i > kq/u$ will grow.

From eqn (13.29) we see that $c_i v_i$ declines with increasing i. For a large number m of strains we can make the approximation $c_m v_m = 0$, which leads to

$$v = \frac{b r_m}{kq - u r_m}. \tag{13.33}$$

This equation describes the relation between virus load, v, and antigenic diversity, m, in patients below the diversity threshold.

13.2.1 Cross-sectional comparisons

Let us first assume that patients differ in their strain-specific responses (parameters p and c), but not in their cross-reactive responses (parameters q and k). If the patients are able to control the virus (inequality (13.32) holds), then eqn (13.33) describes a *negative* correlation between viral load and diversity. If the patients are unable to control the virus (inequality (13.32) does not hold), then the virus population tends to very high values and the number of strains, i, which will be present is given by the inequality $r_i > kq/u$. But in terms of relative frequency some of these strains may converge to zero. The minimum number of strains which are necessary to overcome the diversity threshold is given by the smallest integer m such that inequality (13.32) is violated. Exactly the same situation applies when the only difference among patients is the rate at which individual virus strains replicate.

If, on the other hand, patients differ in their cross-reactive immune response against the virus (parameters q and k), but not in their strain-specific immune response, we find a positive correlation between virus load and diversity. For a large number of strains, we can again make a continuous approximation for the replication rates, $r_m = \rho(m)$. If the individual r_i are uniformly distributed on the interval $(0, R)$, then $\rho(m) = R(1 - m/n)$. As before we find that the number of coexisting strains at equilibrium, m, is the root of the quadratic equation $m^2 + 2\phi m - 2\phi n = 0$ with $\phi = cp/(kq)$ and $c = c_i$ for all strains. Using this equation to eliminate kq in eqn (13.33), we get

$$v = \frac{bRm^2}{2cpn - uRm^2}.$$ (13.34)

This describes a *positive* correlation between viral load and diversity.

13.3   Comparison with data

The advanced models of antigenic variation add important insights to our understanding of viral diversity in patients. If viral strains have different replication rates, there is selection for the faster replicating strain. Without strain-specific immune responses, there would be competitive exclusion: the fastest strain will win, all others go to extinction. Immune responses provide selection pressure for or against antigenic diversity. Strain-specific responses favour diversity, while cross-reactive responses work against diversity. Specifically, we list the following results of the mathematical models discussed in this chapter:

1. In an individual infection, increasing the number m of viral variants (by increasing the total number of variants, n) leads to increasing virus load. Antigenic diversity increases virus load.

2. The immune system can select for or against antigenic diversity. The amount of antigenic diversity which is selected in a given patient depends on the ratio of strain-specific to cross-reactive immune responses: the stronger the strain-specific component of the immune system the more diversity, the stronger the cross-reactive component the less diversity (eqn (13.19)).

3. In cross-sectional comparisons among different patients, the correlation between virus load and antigenic diversity can be positive or negative. If patients differ in their strain-specific responses, then a weak responder allows a high virus load, but also provides little selection for variation, while a strong responder reduces virus load to low levels but selects for high diversity. The resulting relation between load and diversity is negative. If patients differ in their cross-reactive responses against the virus, then the models predict a positive correlation between load and diversity: weak responders allow high virus load, and also provide low selection against antigenic variation.

Studies of genetic variation in HIV infected patients have shown that rapid disease progression can be associated with weak immune responses, high virus load and low genetic diversity. Such a pattern is predicted by the 'antigenic diversity threshold' theory if patients differ mostly in their strain-specific immune responses.

Nevertheless, such findings have sometimes—incorrectly—been interpreted as arguing against the diversity threshold idea. Wolinsky *et al.* (1996) followed six HIV-1 infected patients longitudinally over up to five years after infection. Two patients were rapid progressors and died within 36 and 42 months after infection. Genetic diversity was sampled in a region of the envelope protein. CTL precursor frequency was determined against the Env, Gag, and Pol protein. Wolinsky interpreted his data as rejecting our theory, apparently because high genetic diversity did not correlate with rate of CD4 cell loss and in particular the two patients who progressed rapidly to disease had low genetic diversity.

Wolinsky's data are difficult to interpret. It is unclear whether the observed *genetic variation* represents *antigenic variation*. The CTL data suggest that only one patient has a significant response against Env, while the five other patients have low or undetectable CTL precursors against Env. The major CTL responses in these patients are directed against the Gag and Pol proteins, which have not been sequenced. The genetic variation in Env could certainly include escape from antibody responses, but these responses have not been measured. A precise test of our theory has to investigate antigenic variation in defined immunodominant epitopes. Genetic variation alone need not correlate with CD4 decline or virus load increase.

Even if we were to accept that Wolinsky's genetic variation is somehow indicative of antigenic variation, the data would by no means argue against the validity of the diversity threshold theory, but would be in agreement with the scenario that we proposed years before Wolinsky's study. The claim that such findings refute our theory rests on Wolinsky's incorrect interpretation that the magnitude of the diversity threshold is the same in different patients irrespective of their immune response against HIV. We have always stated the opposite: if patients differ in their immune responses against the virus (and in other parameters), then they will have different antigenic diversity thresholds. In particular, patients with weak immune responses have low diversity threshold values and will progress to disease rapidly and without significant antigenic variation. In contrast, patients with strong strain-specific immune responses have high diversity thresholds and will progress to disease rapidly and with antigenic variation. Such a picture is in fact suggested by Wolinsky's and other studies (see Delwart *et al.*, 1994).

13.4 Further reading

Nowak and May (1992) study coexistence and competition in viral quasispecies. See also Bittner *et al.* (1997) and Nowak and Bangham (1996). Experimental observations on diversity and disease progression are Ljunggren *et al.* (1989), Meyerhans *et al.* (1989), Albert *et al.* (1990), Nara *et al.* (1990), Goudsmit *et al.* (1991), Mullins *et al.* (1991), Wolfs *et al.* (1991), Delassus *et al.* (1992), Kuiken *et al.* (1992), McNearney *et al.* (1992), Zhang *et al.* (1993), Connor and Ho (1994), Delwart *et al.* (1994), Fenyo (1994), Goudsmit (1995), Lukashov *et al.* (1995), Pelletier *et al.* (1995), Ferbas *et al.* (1996), Koot *et al.* (1996), Wolinsky *et al.* (1996), Delwart *et al.* (1997), Ganeshan *et al.* (1997), Ostrowski *et al.* (1998), Shankarappa *et al.* (1999). For a reply to Wolinsky *et al.* (1996) see Nowak *et al.* (1996*a*).

14

MULTIPLE EPITOPES

The previous two chapters described antigenic variation in a single epitope. In this chapter we develop mathematical models of antigenic variation in multiple epitopes. The resulting 'Multiple epitope theory' leads to significant refinements in our understanding of the dynamical interaction between genetically variable pathogens and the immune system.

An epitope is a part of a viral (or other) protein that can be seen by the immune system. Antibodies bind to three-dimensional conformations of proteins. Therefore, an antibody epitope can consists of amino acids located in different parts of the primary structure (sequence). T cells, on the other hand, recognize *linear* epitopes, which consist of about 10 consecutive amino acids. A typical protein contains several epitopes that can be seen by either antibody or T-cell responses. There are two kinds of T-cell epitopes: those that are seen by cytotoxic T cells (CTL) and those that are seen by helper T cells. CTL epitopes are presented by major histocompatibility complex (MHC) class I proteins, whereas helper T-cell epitopes are presented by MHC-II proteins.

While the models in this chapter apply equally to antibody and T-cell immune responses, we focus on CTL responses, because extensive experimental data have been collected on antigenic variation of HIV-1 in CTL epitopes. Pioneering work by Andrew McMichael, Rodney Phillips and colleagues at the Institute of Molecular Medicine in Oxford has characterized CTL immune responses in HIV infection and shown that the virus can generate mutants which escape from CTL recognition.

CTL responses are believed to play an important role in controlling HIV infection. Strong CTL responses arise early in infection and persist for most of the asymptomatic phase. Rapid-progressors tend to have weaker CTL responses than slow-progressors. CTL can recognize and eliminate virus infected cells; they can also release chemicals which inhibit virus entry into cells and/or virus replication within cells. Therefore, escape from CTL responses is likely to be important for HIV persistence and disease progression.

In this chapter, we develop mathematical models for CTL responses against several epitopes. Antigenic variation may occur in each epitope. Even the simplest such models can have very complicated dynamics, with surprising features. There can be complex *antigenic oscillations*, with distinct peaks of different virus variants and fluctuations in the size and specificity of the immune responses. The emergence of an escape mutant in one epitope can shift the immune response to another (weaker) epitope. Our model also provides a general theory for immunodominance in the presence of antigenic variation.

Immunodominance is the frequently observed phenomenon that the immune response against viral (or other) pathogens often focuses against one or a few epitopes, while many epitopes would be present on this pathogen. In our mathematical models, immunodominance arises as a consequence of competition between CTL responses for antigenic stimulation. CTL proliferate in response to antigenic stimulation. A more immunogenic epitope stimulates a stronger response, which in turn down-regulates the virus population to lower levels, and thus reduces the stimulation of other, less immunogenic CTL responses.

The situation is very similar to a classical predator–prey interaction in ecology. Viruses are prey, and CTL are predators. CTL 'eat' virus and multiply proportional to the abundance of virus. Viruses reproduce, but are constantly eliminated by CTL. If different predators feed on the same prey, then usually only one, the most efficient, predator will survive at equilibrium. All other predators will go to extinction. This is known as the *competitive exclusion principle*.

In terms of our model, we find that for an antigenically homogeneous virus population only the most efficient CTL response tends to survive while all others go to very low levels (and possibly disappear). 'Efficiency' is determined by the immunogenicity of the epitope, that is the rate at which this particular epitope stimulates CTL proliferation. In molecular terms, several factors contribute to immunogenicity: a peptide is likely to be highly immunogenic, if it binds the presenting MHC molecule with high affinity, if the resulting MHC–peptide complex is present in abundance on the surface of the infected cell, if this complex has a high affinity for the T-cell receptor, and if the reacting T cells have high precursor frequency. The abundance of an epitope will be influenced by the concentration of the source protein, by its intrinsic stability and localization, and by the presence of appropriate proteolytic cleavage sites that lead to efficient antigen processing.

The multiple epitope theory provides a clear understanding of the events that can follow the emergence of an escape mutant (Fig. 14.1). Consider a homogeneous virus population, and CTL responses against two epitopes, A and B. Suppose epitope A is more immunogenic and therefore elicits the immunodominant response; the response against epitope B is essentially zero. Imagine a mutant arises in epitope A which is not seen by the specific CTL response against the original variant in A. One of four things can happen:

(i) The mutant may induce a new specific CTL response. The virus population is controlled by two specific CTL responses against the two variants in epitope A; there is no response against epitope B.

(ii) The mutant may not elicit a specific immune response to epitope A, but may induce a partial shift in immunodominance to epitope B. The virus population is then controlled by a response against epitope B and a response against the original variant in epitope A; the mutant epitope A is not seen by the immune system.

(iii) The mutant may induce a new specific response in epitope A, which outcompetes the response against the original virus, and in addition induce a partial shift in immunodominance to epitope B. The virus population is then controlled by a response against epitope B and a response against the new variant in epitope A; the response against the original variant in epitope A declines to low levels.

(iv) The mutant may induce a complete shift in immunodominance to epitope B. The response against A declines, and the virus population is only controlled by the response against B.

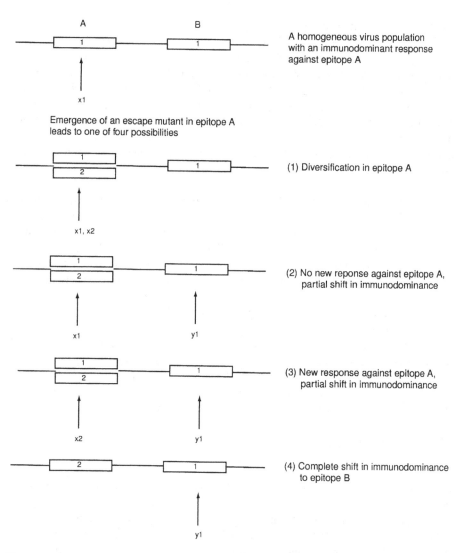

Fig. 14.1 Antigenic variation can shift immunodominance. The emergence of an escape
variant in an immunodominant epitope, A, leads to one of four possible outcomes:
(i) diversification in epitope A leading to a virus population with two epitope variants
in A both of which are seen by specific CTL. (ii) Partial shift of immunodominance
to epitope B, no specific response against the escape variant in A. (iii) Partial shift of
immunodominance to epitope B, together with a change in specificity of the response
to epitope A. (iv) Complete shift of immunodominance to epitope B, together with
a complete replacement of the original variant in A by the escape mutant. Specific
CTL to the original variant in epitope A, the escape variant in epitope A and epitope B
are denoted by x_1, x_2, and y_1, respectively.

Thus, the dynamical events following the emergence of an escape mutant in models with several epitopes are qualitatively different to what could occur in models with a single epitope. In the multiple epitope theory an important new consequence of antigenic variation is that it can shift the dominant immune response to another epitope. Specifically, antigenic variation in the immunodominant epitope can divert the immune responses to less immunogenic epitopes, thereby reducing immune control of the virus. In addition, in multiple epitope models it is possible that escape mutants arise, which never induce a new specific immune response against themselves, but do nevertheless not grow to fixation (that is 100% abundance). In fact, depending on the detailed rate constants such 'unseen escape mutants' can adopt any equilibrium frequency.

Furthermore, the multiple epitope models can give rise to complex *antigenic oscillations* which may occur in the absence of ongoing mutational events. The models generate distinct peaks in viral abundance, which are often dominated by single viral genotypes and occur whenever the responses against particular variants have declined to low levels because of lack of stimulation. For these antigenic oscillations to occur, it is not essential that mutation continuously generates new antigenic material. Antigenic oscillations are different to the previously described 'antigenic drift', where the emergence of a new variant gives rise to a peak in viral abundance. Antigenic oscillations arise as a consequence of the nonlinear dynamics of the immune responses acting on existing viral diversity.

For HIV infection, the multiple epitope theory suggests that antigenic variation and as a consequence shifting of the immune response to weaker epitopes may be an important route of disease progression. Initially the anti-HIV CTL response may be focused against highly immunogenic epitopes, but antigenic variation in these epitopes may divert the CTL response to less immunogenic epitopes and thereby reduce the overall efficiency of the CTL response and increase virus load. Increasing virus load means disease progression.

Contrary to intuition, the models suggest that those patients that have a stable CTL response against a conserved epitope control the virus more efficiently than patients that have fluctuating responses against several epitopes. Long-term recognition of several epitopes is, in terms of our model, a consequence of antigenic variation and, therefore, implies reduced immune control. Immunotherapy against HIV shall aim to boost responses to conserved epitopes rather than the most immunogenic epitopes, which may be variable. The picture is somewhat complicated by the very recent theoretical observation that stable recognition of several epitopes can also be the consequence of CTL *memory* responses (Wodarz and Nowak 2000).

In Section 14.1 we review experimental evidence of antigenic variation in HIV.

In Section 14.2 we develop the simplest 'multiple epitope' model, in which all virus mutants have the same overall replication rates and all variants of a given epitope have the same immunogenicity (different epitopes can, however, have different immunogenicities). CTL responses proliferate proportional to the product of their own abundance and (specific) viral abundance.

In Section 14.3 we allow different replication rates and immunogenicities for the different virus mutants. The equations of Sections 14.2 and 14.3 are specific Lotka–Volterra systems.

Throughout the chapter we analyse models with two epitopes. There may be n_1 variants in epitope A and n_2 variants in epitope B. This leads to a total of $n_1 \times n_2$ potential virus mutants. Together with n_1 specific CTL clones against epitope A and n_2 specific CTL clones against epitope B, we have a system with $n_1 \times n_2 + n_1 + n_2$ dimensions. We do not consider models with more than two epitopes, but the extension to such models is possible.

In general we cannot give a complete analysis of the dynamical possibilities for situations where different virus mutants have different replication rates. Therefore, we give complete classifications of some low dimensional cases. In Section 14.4 we discuss the 2×1 case (i.e. $n_1 = 2$, $n_2 = 1$), which provides a complete description of the dynamical events after the emergence of an escape mutant in a homogeneous virus population. We will see how antigenic variation can shift immunodominance.

In Section 14.5 we examine the effect of cross-reactive immune responses within an epitope. In Section 14.6 we include intracellular competition for MHC presentation. In Section 14.7 we discuss some consequences for immunotherapy. Section 14.8 gives a summary of the biological implications of this chapter.

14.1 Experimental evidence

The formulation of the multiple epitope theory was motivated by detailed studies of anti-HIV CTL responses by Phillips *et al.* (1991). This work describes an extensive longitudinal study of CTL responses against epitopes in the HIV gag protein in six HIV infected patients. Three of these patients recognized gag through HLA B27 (HLA stands for human leucocyte antigen and is the human MHC). They recognized only a single epitope, which remained conserved throughout the study. The three other patients recognized three different epitopes in gag restricted by HLA B8. The virus population of these patients showed genetic variation in these epitopes, and some virus mutants were not recognized by the patients' CTL response. Furthermore, there were unexpected fluctuations in the specificity of the CTL responses. At different time points responses against different epitopes were predominant. Interestingly, the three patients with oscillating responses seemed to develop AIDS faster than the three patients with constant responses.

Figure 14.2 shows data obtained from patient 020 who has an HLA B8 restricted CTL response against the virus. In July 1989, patient 020 recognized epitope A (peptide sequence p24-13) in the gag p24 protein. In December 1989 the same patient had CTL activity against epitope B (peptide p24-20), but no longer against A. In February 1990, there was hardly any response against A or B, instead the CTL response recognized another epitope, C (peptide sequence p17-3 from p17 gag). During 1990 there were fluctuating responses against these three epitopes. Throughout 1989 and 1990 there was extensive genetic variation in the epitopes A and C, although epitope B was rather conserved. In all three epitopes there were variants which were at certain time points not seen by the patient's CTL. There is some evidence for selective increase of such escape variants: the E variant in epitope A increased in frequency from 0% to 100% at a time when specific CTL failed to recognize this epitope. Similarly, a poor CTL response to the K variant in epitope C is associated with an accumulation of this variant. But overall

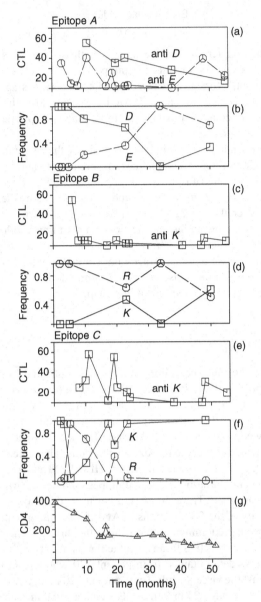

Fig. 14.2 Patient 020 has responses to three HLA-B8-restricted epitopes in gag, designated *A*, *B* and *C*. (a) The strength of the CTL response is measured in % specific lysis of HLA-matched target cells that carry the relevant epitope. The patient's CTL fluctuate in their capacity to recognize the two variants of epitope *A*, GEIYKRWII (*E* variant) and GDIYKRWII (*D* variant). Both peptides bind HLA B8 with similar efficiency. (b) Changes in frequency of the *D* and *E* variants of epitope *A* in proviral DNA. Initially only the *D* variant was detected in proviral DNA. Loss of the CTL response against the *E* variant at around 20 months was associated with its emergence as the only detectable variant. This appears to be an example of escape from a CTL

the pattern is complex, and we observe changes in the frequency of variant viruses with shifting immunodominance in the CTL responses.

Figure 14.3 shows data from patient 007, a slow progressor who remains well and has a sustained CTL response to a single HLA-B27-restricted *gag* epitope. In this epitope there is one predominant and one rare variant, but both variants are seen by the patient's CTL.

Borrow *et al.* (1997) describe a detailed study of antigenic variation during the primary infection of a patient. Initially the patient has a strong CTL response against a particular epitope in the Env protein. There is rapid emergence of an escape mutant which carries a single-point mutation that prevents recognition by the patient's CTL. Within about 100 days there is complete replacement of the wild-type virus by the escape mutant. At this time the patient has mounted several new CTL responses against other epitopes in Env, Gag, Pol and Nef. Hence the general picture is in agreement with our notion that escape in the immunodominant epitope can shift the CTL response to other epitopes.

14.2 The simplest multiple epitope model

First we will develop a simple model that describes the dynamics of immune responses against two different epitopes:

$$
\begin{aligned}
\dot{v}_{ij} &= v_{ij}(r - px_i - qy_j), \\
\dot{x}_i &= x_i(cv_{i*} - b) \quad \text{with } i = 1, \ldots, n_1, \\
\dot{y}_j &= y_j(kv_{*j} - b) \quad \text{with } j = 1, \ldots, n_2.
\end{aligned}
\tag{14.1}
$$

response at a time when the circulating CTL clones fail to recognize the variant; later a cross-reactive response returned and the selection pressure altered. (c) CTL response to epitope *B*, which must be considered in conjunction with the response to epitope *A*, 60 amino acids upstream in HIV Gag. The CTL response against the *K* variant was initially present and then lost. (d) Frequency of variants in epitope *B*. The peptide DCRTILKAL (*R* variant), which cannot be recognized by the patient's CTL although it binds to HLA B8, predominated. Later, the recognizable variant, DCKTILKAL (*K* variant), dominated but there were no specific CTL detectable. The linkage of variants in epitopes *A* and *B* changed: at 23 months the *D* + *R* variants were linked, while at 34 months the *E* + *R* variants were the majority species. (e) The CTL response against epitope *C* was variable as the predominant sequence shifted from GGKKKYKLK (*K* variant) to GGRKKYKLK (*R* variant) and back. CTL never recognised the *R* variant, though it binds to HLA B8. (f) Changing frequency of the variants in epitope *C*. The *K* variant became more abundant when the CTL response to this epitope was low. (g) Serial CD4 counts. Patient 020 had a progressive loss of CD4 T cells, developing AIDS with a CD4 T-cell count of $< 200 \, \mu \mathrm{l}^{-1}$, and *Pneumocystis carinii* pneumonia. He died of a gastrointestinal haemorrhage.

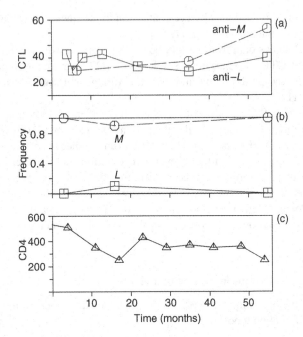

Fig. 14.3 Patient 007 has a sustained response to an HLA-B27-restricted epitope in gag. (a) Percentage of specific lysis of HLA-matched target cells treated with the epitope decamers KRWIIMGNK (*M* variant) or KRWIILGNK (*L* variant). (b) Sequencing of proviral DNA revealed that virtually all virus in PBMC consists of the *M* variant rather than the *R* variant which represents the data-base consensus sequence. (c) Patient 007 is a fairly slow progressor with a slow decline in CD4 T-cell count to 300 cells $\mu\,l^{-1}$.

Here v_{ij} denotes the abundance of virus variants with sequence i in epitope A and sequence j in epitope B. There are n_1 different sequences for epitope A and n_2 for epitope B. Thus in total we consider $n_1 \times n_2$ different virus variants. The variables x_i and y_j denote CTLs directed at sequence i of epitope A and sequence j of epitope B, respectively. There are n_1 CTL clones directed at the various A variants and n_2 against the B variants. In this simplest model, all the virus variants reproduce at rate r. They are killed by CTL responses at the rates $-pv_{ij}x_i$ and $-qv_{ij}y_j$. CTLs are stimulated by their specific epitope sequence (in association with HLA presentation) and replicate at the rates $cx_i v_{i*}$ and $ky_j v_{*j}$, with the notation

$$v_{i*} = \sum_{j=1}^{n_2} v_{ij} \quad \text{and} \quad v_{*j} = \sum_{i=1}^{n_1} v_{ij}. \tag{14.2}$$

Thus, a particular CTL clone recognizes all viruses that have the specific sequence in the right epitope, i.e. x_i is directed at (and is stimulated by) $v_{i1} + v_{i2} + \cdots + v_{in_2} = v_{i*}$, whereas y_j recognizes $v_{1j} + v_{2j} + \cdots + v_{n_1 j} = v_{*j}$. The constants c and k describe the immunogenicities of the two epitopes. If $c > k$ then epitope A is more immunogenic and will provide a stronger stimulus for replication of the relevant CTL clones. Finally, we assume that in the absence of antigenic stimuli the activated CTLs decline at the rates $-bx_i$ and $-by_j$. In total the model has seven parameters (r, p, b, c, k, n_1 and n_2) and $n_1 n_2 + n_1 + n_2$ variables (dimensions).

The above model assumes a very simple dynamics for the immune responses. We will first analyse this simple model and then study how alterations in the dynamics of the immune response will affect the outcome.

We start by looking for equilibrium solutions. Setting $\dot{x}_i = 0$ we obtain the non-trivial solution $v_{i*} = b/c$ (alternatively, $x_i = 0$). For the total virus abundance this yields the equilibrium value $v = bn_1/c$. From $\dot{y}_i = 0$ we similarly get $v_{*j} = b/k$ and hence $v = bn_2/k$. Both relations cannot be fulfilled (as long as $n_1/c \neq n_2/k$, which is the generic assumption). Hence, there is no interior equilibrium of system (14.1). The competition between the responses against the two different epitopes is decided by the relative magnitudes of the ratios c/n_1 and k/n_2. If, for example, $c/n_1 > k/n_2$ then all y_j converge to zero and the system (14.1) reduces to

$$
\begin{aligned}
\dot{v}_{i*} &= v_{i*}(r - px_i), \\
\dot{x}_i &= x_i(cv_{i*} - b).
\end{aligned}
\tag{14.3}
$$

This is a simple Lotka–Volterra system with neutral oscillations around $v_{i*} = b/c$ and $x_i = r/p$. The eigenvalues of the Jacobian matrix are given by $\pm i\sqrt{rb}$, and hence the period of the oscillations is roughly $T \approx 2\pi/\sqrt{rb}$ (lengthening in the usual way for oscillations of larger amplitudes). The total amount of virus and of immune cells fluctuate indefinitely around their long-term averages $v = bn_1/c$ and $x = n_1r/p$. Note that both quantities increase linearly with the number of variants n_1. Such neutral oscillations are structurally unstable, and therefore we will study structural modifications of the above model in the subsequent sections.

The model also has an interesting kind of degeneracy: the long-term averages of v_{i*} are exactly specified, but the individual v_{ij} remain undefined (within the limits set by v_{i*}). For our entirely deterministic model this means the variants in epitope B are fixed in some arbitrary population structure (satisfying $v_{i*} = b/c$), once the immune responses against B have vanished. In a real situation genetic drift will follow.

Figure 14.4 illustrates the dynamics of eqn (14.1) for $n_1 = n_2 = 2$. Thus there are two variant sequences in each epitope, which induce two x- and two y-responses. All mutants are present in the beginning ($t = 0$); there is no production of new mutants over the time of the simulation. Nevertheless, the system displays antigenic oscillations. The y-responses will slowly converge to zero, leaving the $x_i - v_{i*}$-system in neutral oscillations. Antigenic oscillations can occur without the emergence of new mutants.

14.3 Different parameters for different mutants

In the above model we assumed that all virus mutants have the same replication rates, are killed by CTL responses at equal rates, and induce CTL responses at equal rates. We shall now generalize these assumptions.

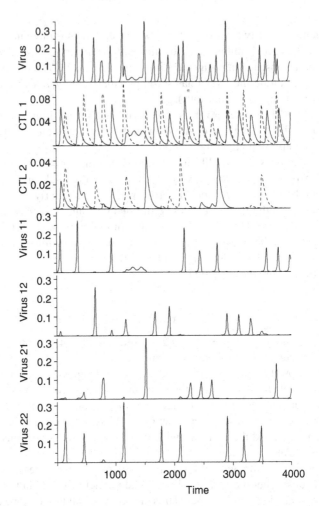

Fig. 14.4 Computer simulation of the basic model given by eqn (14.1). There are two different epitopes with two sequence variants in each epitope (thus altogether two different virus species). All virus mutants replicate at the same rate and are present at the beginning ($t = 0$) in different abundances. There is no subsequent production of new antigenic material. Nevertheless, we observe sequential peaks in viral abundance that correspond to antigenically different variants. Thus antigenic oscillations can occur without antigenic drift. As discussed in the text, we have a clear understanding of the long-term behaviour of the system: the y_i will converge to zero, and there are undamped neutral oscillations with the x_i and v_{i*}. The parameters are: $n_1 = n_2 = 2$, $r = 0.1, p = 5, c = 1.1, k = 1, b = 0.02$. The time axes is in arbitrary units (but the biological observations suggest these oscillations to occur on a time-scale of weeks or months).

Let us first consider the system where the virus mutants differ in their immunological parameters, but still have the same replication rates

$$
\begin{aligned}
\dot{v}_{ij} &= v_{ij}(r - p_i x_i - q_j y_j), \\
\dot{x}_i &= x_i(c_i v_{i*} - b) \quad \text{with } i = 1, \ldots, n_1, \\
\dot{y}_j &= y_j(k_j v_{*j} - b) \quad \text{with } j = 1, \ldots, n_2.
\end{aligned}
\tag{14.4}
$$

The immunogenicity of sequence i in epitope A is now given by c_i, and similarly k_j for j in epitope B. We still maintain that the natural death/decay rate of CTLs, b, is the same for all different specificities. Biologically this seems plausible, but we stress that it is not an essential assumption for the mathematical analysis.

A simple rescaling, $x'_i = p_i x_i$ and $y'_j = q_j y_j$, shows that we can neglect the parameters p_i and q_j. Thus, without loss of generality we can write (after dropping the primes)

$$
\begin{aligned}
\dot{v}_{ij} &= v_{ij}(r - x_i - y_j), \\
\dot{x}_i &= x_i(c_i v_{i*} - b) \quad \text{with } i = 1, \ldots, n_1, \\
\dot{y}_j &= y_j(k_j v_{*j} - b) \quad \text{with } j = 1, \ldots, n_2.
\end{aligned}
\tag{14.5}
$$

Generically, there is again no interior equilibrium, but competitive exclusion between the x- and y-responses. In fact there are only two possible (non-trivial) equilibria: either all x_i converge to zero or all y_j converge to zero. The x-responses will eventually win if

$$
\sum_{i=1}^{n_1} \frac{1}{c_i} < \sum_{j=1}^{n_2} \frac{1}{k_j}.
\tag{14.6}
$$

This can be seen by considering the products

$$
P = \prod_{i=1}^{n_1} x_i^{1/c_i} \quad \text{and} \quad Q = \prod_{j=1}^{n_2} y_j^{1/k_j}.
\tag{14.7}
$$

We obtain

$$
\frac{d(P/Q)}{dt} = \frac{P}{Q}\left(\sum_{j=1}^{n_2} \frac{1}{k_j} - \sum_{i=1}^{n_1} \frac{1}{c_i} \right) b.
\tag{14.8}
$$

Thus, if inequality (14.6) holds, the ratio P/Q will be an exponentially increasing function of time, suggesting that the x-response outperforms the y-response. P/Q is a Lyapunov function. There is asymptotic convergence to the boundary where at least one y_j is zero. For specificity, let us assume that y_1 converges to zero. What happens next?

We can show that it is not possible to have an equilibrium in the face $\{y_1 = 0\}$ with some other y_j and v_{*1} being positive. This can be seen by considering the ratios v_{ih}/v_{i1}

for different i and h. (Since $v_{*1} > 0$ we can always find at least one $v_{i1} > 0$.) With $y_1 = 0$ we get

$$\frac{\mathrm{d}}{\mathrm{d}t}\left(\frac{v_{ih}}{v_{i1}}\right) = -\frac{v_{ih}}{v_{i1}}y_h. \tag{14.9}$$

Thus, for $y_h > 0$ we have that v_{i1} always grows faster than v_{ih} except if $v_{ih} = 0$. But at least some v_{ih} have to be strictly positive. If $v_{*h} = 0$ then y_h converges to 0, and hence there is indeed no equilibrium with some y_j being positive.

If all y_j have vanished, we are left with the system

$$\dot{v}_{i*} = v_{i*}(r - x_i),$$
$$\dot{x}_i = x_i(c_i v_{i*} - b) \quad \text{with } i = 1, \dots, n_1. \tag{14.10}$$

This represents, essentially, n_1 uncoupled oscillators. Note also that now $\dot{v}_{ij} = v_{ij}(r - x_i)$ for all j, whence the ratios v_{ij}/v_{ik} remain constant. As before x_i has neutral oscillations around the equilibrium (time-average) r/p_i and v_{i*} oscillates around b/c_i. The period of the oscillation close to the equilibrium is approximately $T \approx 2\pi/\sqrt{rb}$.

More difficult to understand is the general system with different replication rates for different mutants. Let us assume that the virus mutant v_{ij} replicates with rate r_{ij}. We have

$$\dot{v}_{ij} = v_{ij}(r_{ij} - x_i - y_j),$$
$$\dot{x}_i = x_i(c_i v_{i*} - b) \quad \text{with } i = 1, \dots, n_1, \tag{14.11}$$
$$\dot{y}_j = y_j(k_j v_{*j} - b) \quad \text{with } j = 1, \dots, n_2.$$

Equation (14.8) remains true. Let us again assume that inequality (14.6) holds. Then the ratio P/Q is exponentially increasing over time, implying convergence to a boundary where at least one y_j is zero. Again let us assume that $y_1 = 0$. Now it is indeed possible to have an equilibrium in the face $\{y_1 = 0\}$ with all other x_i and y_j being positive. From $\dot{x}_i = 0$ for all $i = 1, \dots, n_1$ we get $v = b\sum_{i=1}^{n_1} 1/c_i$. From $\dot{y}_j = 0$ for all $j = 2, \dots, n_2$ we get $v = v_{*1} + b\sum_{j=2}^{n_2} 1/k_j$. Thus, an equilibrium in the interior of the face $\{y_1 = 0\}$ is possible provided

$$v_{*1} = b\left(\sum_{i=1}^{n_1}\frac{1}{c_i} - \sum_{j=2}^{n_2}\frac{1}{k_j}\right) > 0. \tag{14.12}$$

Hence, combining this inequality with the earlier (14.6), we have

$$\sum_{j=1}^{n_2}\frac{1}{k_j} > \sum_{i=1}^{n_1}\frac{1}{c_i} > \sum_{j=2}^{n_2}\frac{1}{k_j} \tag{14.13}$$

as a necessary condition for the existence of such an equilibrium. We do not know any sufficient condition, but for the ratios v_{ih}/v_{i1} we get

$$\frac{\mathrm{d}}{\mathrm{d}t}\left(\frac{v_{ih}}{v_{i1}}\right) = \frac{v_{ih}}{v_{i1}}(r_{ih} - r_{i1} - y_h). \tag{14.14}$$

Thus, an equilibrium is only possible if $r_{ih} > r_{i1}$ for those indices i and h where $v_{ih} > 0$.

We lack a complete understanding of system (14.11), but conjecture the following dynamics. Suppose inequality (14.6) holds. Then one y_j will converge to zero (say y_1). If $\sum_{i=1}^{n_1} 1/c_i < \sum_{j=2}^{n_2} 1/k_j$, then another y_j will converge to zero. Several y_j will become extinct until the inequality reverses, e.g. $\sum_{i=1}^{n_1} 1/c_i > \sum_{j=l}^{n_2} 1/k_j$. (Thus we have assumed that all $y_j =$ with $j = 1, \ldots, l - 1$ have converged to zero.) Now a coexistence between the remaining y_j and the x_i is possible, but depends on the r_{ij} (with $j = 1, \ldots, l - 1$) being small compared to the replication rates of other mutants that are present at this equilibrium—as specified by eqn (14.14). The right-hand side of eqn (14.14) can only be zero if $r_{ih} > r_{i1}$.

Other complications are possible. Suppose, by chance, one of the x_i dies out first. Then the y-responses may become immunodominant. Thus, clonal exhaustion can indeed lead to switching immunodominance. And which epitope is immunodominant in the long run is partly determined by the random events of extinction of CTL clones at very low frequencies.

Appendix B.1 contains a general result. Any fixed point of (14.11) is neutrally stable within its face, and a generalization of Volterra's function represents an invariant of motion.

Another biologically significant observation is that *either* some x_i or some y_j become extinct, but never both. Let us consider a situation where a specific x_i and a specific y_j have become extinct. Then $\dot{v}_{ij} = r_{ij}v_{ij} > 0$, and v_{ij} is able to invade this fixed point. Hence such a fixed point cannot be saturated (i.e. stable against invasion by those variables which are close to zero).

14.3.1   *What determines immunodominance?*

Even for system (14.11), where the virus mutants have different replication rates, it can happen that all responses against one epitope will vanish, and the whole immune response is entirely directed against the other epitope. We will call this state 'complete immunodominance'. What are the conditions for complete immunodominance?

We have seen that the inequality (14.6),

$$\sum_{i=1}^{n_1} \frac{1}{c_i} < \sum_{j=1}^{n_2} \frac{1}{k_j}, \tag{14.15}$$

implies that at least one y_j will tend towards zero. Furthermore, we also know that for a saturated equilibrium *either* some x_i- or some y_j-responses can vanish, but never both. Thus inequality (14.6) is a necessary condition for complete immunodominance of the x-response.

If all r_{ij} are the same then inequality (14.6) is sufficient for complete immunodominance. The epitope which minimizes the sum of the reciprocals of the immunogenicities of all variants is immunodominant. The responses against all other epitopes will vanish.

If the r_{ij} are different then complete immunodominance would imply that we are left with the subsystem

$$
\begin{aligned}
\dot{v}_{ij} &= v_{ij}(r_{ij} - x_i), \\
\dot{x}_i &= x_i(c_i v_{i*} - b).
\end{aligned}
\tag{14.16}
$$

For each i, only one v_{ij} will persist. This will be the v_{ij} with the largest r_{ij} in this row of the r_{ij}-matrix. More precisely let us define the index m_i as $r_{im_i} = \max_j\{r_{ij}\}$. The equilibrium of (14.15) is given by

$$
x_i = r_{im_i}, \quad v_{i*} = v_{im_i} = \frac{b}{c_i}.
\tag{14.17}
$$

This equilibrium is saturated (as defined above) with respect to the y_j-responses if $\partial \dot{y}_j / \partial y_j = k_j v_{*j} - b < 0$. We have at equilibrium

$$
v_{*j} = \sum_{i=1}^{n_1} \delta(m_i, j) v_{im_i} = b \sum_{i=1}^{n_1} \frac{\delta(m_i, j)}{c_i},
\tag{14.18}
$$

where δ is the Kronecker symbol, i.e. $\delta(m_i, j) = 1$ if $m_i = j$ and $\delta(m_i, j) = 0$ otherwise. Thus saturation requires

$$
\sum_{i=1}^{n_1} \delta(m_i, j) \frac{1}{c_i} < \frac{1}{k_j}, \quad \text{for all } j.
\tag{14.19}
$$

This is a necessary and sufficient condition for complete immunodominance of the x-responses.

14.4 Activated CTLs arise from inactivated precursors

In this section we include the biologically essential assumption that the CTLs are not only produced by replication of already activated cells, but are also generated by activation of specific precursor cells at rates proportional to the specific antigen abundance. This leads to

$$
\begin{aligned}
\dot{v}_{ij} &= v_{ij}(r_{ij} - p_i x_i - q_j y_j), \\
\dot{x}_i &= \eta c_i v_{i*} + x_i(c_i v_{i*} - b) \quad \text{with } i = 1, \ldots, n_1, \\
\dot{y}_j &= \eta k_j v_{*j} + y_j(k_j v_{*j} - b) \quad \text{with } j = 1, \ldots, n_2.
\end{aligned}
\tag{14.20}
$$

A sensible assumption is that the activation signals for precursor cells and already activated cells depend on the immunogenicity of the epitope sequence and are proportional to each other, with η being the proportionality constant. Again this is just one way of writing things more neatly; the assumption of a common η-value is not essential for the following analysis.

Since η is positive it is clear that the x- and y-responses can coexist. An interesting problem arises at once: what are the possible equilibria of system (14.20)? Can some v_{ij} converge to zero? Are there interior equilibria?

Let us first perform the helpful transformation $x'_i = p_i x_i$ and $y'_j = q_j y_j$. This produces (where we have at once dropped the primes):

$$\dot{v}_{ij} = v_{ij}(r_{ij} - x_i - y_j),$$
$$\dot{x}_i = \eta c_i p_i v_{i*} + x_i(c_i v_{i*} - b) \quad \text{with } i = 1, \dots, n_1, \qquad (14.21)$$
$$\dot{y}_j = \eta k_j q_j v_{*j} + y_j(k_j v_{*j} - b) \quad \text{with } j = 1, \dots, n_2.$$

14.4.1 The neutral case: all mutants have the same replication rates

An important special case arises when all virus mutants replicate at the same rate, i.e. $r_{ij} = r$. Without loss of generality we then set $r = 1$. Different epitope sequences may have different immunogenicities and different rates at which they are recognized by the CTL response, but all mutations in these epitopes are essentially neutral with respect to the replication rate.

For the interior equilibria of this neutral system (with $r_{ij} = 1$) we obtain the relations

$$1 = x_i + y_j, \quad \text{for all } i, j. \qquad (14.22)$$

This immediately implies that all x_i (and all y_j) have to be the same, i.e.

$$x_i = \xi \quad \text{and} \quad y_j = 1 - \xi, \quad \text{for all } i, j. \qquad (14.23)$$

For the virus population we then obtain, at equilibrium,

$$v_{i*} = \frac{b\xi}{c_i(\eta p_i + \xi)}, \quad v_{*j} = \frac{b(1 - \xi)}{k_j(\eta q_j + 1 - \xi)}. \qquad (14.24)$$

This specifies the set of all interior equilibria. The constant ξ is obtained from $\sum_i v_{i*} = \sum_j v_{*j}$, which can be a high order polynomial. If all p_i are the same and all q_i are the same, then this relation reduces to a quadratic equation in ξ. Equation (14.24) specifies the equilibrium values for v_{i*} and v_{*j}. The individual v_{ij} can take arbitrary values within this envelope. Essentially there are $n_1 + n_2$ constraining relations for $n_1 \times n_2$ variables. Appendix B.2 shows local stability of this set of equilibria (at least for some special cases).

There are also boundary equilibria (with some $v_{ij} = 0$), but we will show that none of these can be saturated (i.e. stable to invasion by the v_{ij} in question). First note that in each row (or column) at least one v_{ij} has to be positive at a saturated equilibrium point. Otherwise the corresponding x_i or y_j would vanish and at least one v_{ij} of this row (or column) can invade. In general this invasion is specified by the transversal eigenvalue $\lambda_{ij} = \partial \dot{v}_{ij}/\partial v_{ij} = 1 - x_i - y_j$. Let us consider an equilibrium where the x_i and y_j take some arbitrary values. But note that for each x_i there is at least one y_j such that $x_i + y_j = 1$. This follows from the fact that in each row (or column) at least one v_{ij} has to be positive. Similarly for each y_j there is at least one x_i such that $y_j + x_i = 1$. Let us denote by x_1 the smallest of all x_i, and let us denote by y_1 the smallest of all y_j. Clearly $x_1 + y_1 \leq 1$. We have to distinguish two cases:

(i) If $x_1 + y_1 = 1$ then it follows that all x_i are equal to some constant ξ and all y_j are equal to $1 - \xi$. But in this case for any $v_{ij} = 0$ the eigenvalue λ_{ij} has to be zero, hence no saturation (which requires $\lambda_{ij} < 0$).

(ii) If $x_1 + y_1 < 1$ then $v_{11} = 0$ otherwise we are not at an equilibrium. But $x_1 + y_1 < 1$ implies $\lambda_{11} > 0$ and hence no saturation.

Thus there is no saturated equilibrium at the boundary. Putting this fact together with the evidence of local stability (Appendix B.2 and the numerical simulations), we conjecture that all trajectories converge to the manifold of interior equilibria specified by eqns (14.23) and (14.24).

Figure 14.5 gives a numerical example of such a slow convergence towards an interior equilibrium. We chose the same system as for Fig. 14.4, but included the immigration term of eqn (14.20) with $\eta = 0.001$. Again there are antigenic oscillations over a long period, without the emergence of new mutants.

14.4.2   The mutants have different replication rates

We do not have a complete understanding for the case where the r_{ij} can take arbitrary (positive) values. The numerical simulations suggest that all trajectories converge to fixed points, which are generically on the boundary. But we cannot rule out the existence of cyclic solutions or chaotic attractors.

There is an interesting *exclusion principle*. For simplicity consider a case with two epitopes and two sequence variants in each epitope. The conditions for an interior equilibrium are

$$r_{11} = x_1 + y_1, \quad r_{12} = x_1 + y_2,$$
$$r_{21} = x_2 + y_1, \quad r_{22} = x_2 + y_2. \tag{14.25}$$

This can be fulfilled if and only if $r_{11} - r_{12} = r_{21} - r_{22}$, as is the case if mutations in one epitope have no effect on replication rates or if the contributions of the mutations in different epitopes are additive. But this relation will not hold for an arbitrary choice of parameters r_{ij} (except for a set with measure zero). Hence, in general it is not possible to have an interior equilibrium, and at least one of the v_{ij} has to be zero. This means that for a virus population expressing simultaneously two different variants at two different epitopes it is not possible to obtain a stable *selection* equilibrium with all four virus variants present. The generalization of this result to m epitopes and n sequences in each epitope is obvious: for a generic choice of the r_{ij}, all but $(m+n-1)$ of the v_{ij} must vanish. Of course, for a fast mutating virus like HIV, such variants can be maintained at finite values in a mutation equilibrium (facilitated by the additional effect of recombination).

Figure 14.6 shows a computer simulation of eqn (14.20) for $n_1 = n_2 = 3$. The replication rates and immunogenicities of the individual virus variants are randomly

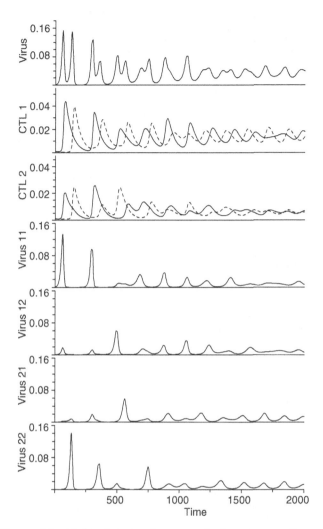

Fig. 14.5 The model with immigration from a pool of precursor cells as given by eqn (14.20). All virus mutants have the same replication rates, $r_{ij} = r$. As shown in Section 14.4.1 there is convergence to a degenerate set of interior equilibria, given by eqns (14.23) and (14.24). The oscillations are damped on a very slow time-scale. The parameters are the same as in Fig. 14.4: $n_1 = n_2 = 2, r = 0.1, p = 5, c = 1.1,$ $k = 1, b = 0.02$; except $\eta = 0.001$.

assigned. We observe damped antigenic oscillations to a boundary equilibrium with only three virus species surviving. These are v_{12}, v_{21}, and v_{33}. Note that there is no cross-reactivity between these three mutants. It seems that the virus population often adopts such a structure. In this example all x_i- and y_j-responses coexist.

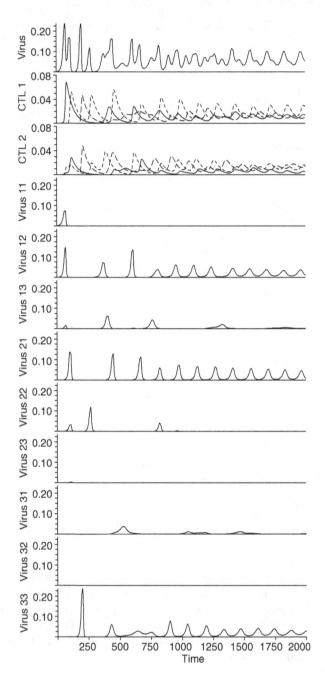

Fig. 14.6 An example of the dynamics of a higher dimensional system with randomly assigned parameters. The figure shows a computer simulation of eqn (14.20) with $n_1 = n_2 = 3$. The randomly assigned parameters are: $r_{11} = 0.0739$, $r_{12} = 0.1175$, $r_{13} = 0.1170$, $r_{21} = 0.1261$, $r_{22} = 0.1478$, $r_{23} = 0.1217$, $r_{31} = 0.0925$, $r_{32} = 0.0738$, $r_{33} = 0.1256$, $p_1 = 5.1361$, $p_2 = 5.2294$, $p_3 = 5.3526$,

14.4.3 *The limit of large η*

In the limit of large η we obtain the system

$$
\begin{aligned}
\dot{v}_{ij} &= v_{ij}(r_{ij} - x_i - y_j), \\
\dot{x}_i &= \eta c_i p_i v_{i*} - b x_i \quad \text{with } i = 1, \dots, n_1, \\
\dot{y}_j &= \eta k_j q_j v_{*j} - b y_j \quad \text{with } j = 1, \dots, n_2.
\end{aligned}
\tag{14.26}
$$

This is derived from eqn (14.20), by neglecting x_i compared to ηp_i in $\dot{x}_i = c_i v_{i*}(\eta p_i + x_i) - b x_i$. In the same way we derive the equation for the y_j.

If all virus mutants have the same replication rates, i.e. $r_{ij} = r$, then system (14.26) converges to a set of interior equilibria, which is given by

$$
x_i = \xi, \quad y_j = r - \xi, \quad v_{i*} = \frac{b}{\eta c_i p_i} \xi, \quad v_{*j} = \frac{b}{\eta k_j q_j}(r - \xi).
\tag{14.27}
$$

From $\sum_{i=1}^{n_1} v_{i*} = \sum_{j=1}^{n_2} v_{*j}$ we obtain

$$
\xi = r \sum_{j=1}^{n_2} \frac{1}{k_j q_j} \bigg/ \left(\sum_{i=1}^{n_1} \frac{1}{c_i p_i} + \sum_{j=1}^{n_2} \frac{1}{k_j q_j} \right).
\tag{14.28}
$$

There are no saturated equilibria on the boundary. The proof for this is equivalent to the one in Section 14.4.2.

For the system with arbitrary replication rates, r_{ij}, we cannot give a complete analysis. The same exclusion principle as in Section 14.4.2 applies. Thus in general there are no interior equilibria. In Appendix B.3 we give a complete classification of all dynamical possibilities for the system $n_1 = n_2 = 2$.

14.5 The 2 × 1 case

14.5.1 $\eta = 0$

Since we cannot derive a general analysis for the case with different replication rates r_{ij}, we will now describe some low dimensional cases. Let us first consider a system with two mutants in epitope A and only a single variant in epitope B, i.e. $n_1 = 2$ and $n_2 = 1$.

$q_1 = 4.6811$, $q_2 = 4.9450$, $q_3 = 4.6972$, $c_1 = 0.8719$, $c_2 = 1.0150$, $c_3 = 0.8753$, $k_1 = 0.7797$, $k_2 = 1.0080$, $k_3 = 0.8708$. In addition we chose $b = 0.02$ and $\eta = 0.001$. There are slowly damped oscillations to a boundary fixed point with all virus mutants zero, except v_{12}, v_{21}, and v_{33}. Note that this three virus mutants do not share a single epitope. We often observe that the virus population converges to such a state with minimal (zero) cross-reactivity.

For $\eta = 0$ we have

$$
\begin{aligned}
\dot{v}_1 &= v_1(r_1 - x_1 - y), \\
\dot{v}_2 &= v_2(r_2 - x_2 - y), \\
\dot{x}_1 &= x_1(c_1 v_1 - b), \\
\dot{x}_2 &= x_2(c_2 v_2 - b), \\
\dot{y} &= y[k(v_1 + v_2) - b].
\end{aligned}
\tag{14.29}
$$

We have avoided unnecessary indices by setting $v_1 = v_{11}, v_2 = v_{21}, r_1 = r_{11}, r_2 = r_{21}$, and $y = y_1$. For system (14.29) we can distinguish five parameter regions, which are mutually exclusive and cover the whole parameter space. For each parameter region there is exactly one saturated equilibrium. All trajectories from the interior of the phase-space converge to the face that contains the saturated equilibrium, and within this face there are neutral oscillations around the equilibrium. The five equilibria, P_1 to P_5, and their conditions of existence and stability are listed in Table 14.1.

We can now understand the dynamics following the emergence of a new mutant in a homogeneous virus population. Let us consider a virus population with only one type of virus; i.e. $v_2 = 0$. There is no coexistence between the two immune responses. Let us assume that $c_1 > k$ (i.e. $1/c_1 < 1/k$). Thus $x_1 = r_1$ and $y = 0$. A mutation in epitope A, i.e. the emergence of mutant v_2, can lead to four different possibilities:

(i) It can simply lead to a diversification in epitope A without stimulating an immune response against epitope B. This happens if $1/c_1 + 1/c_2 < 1/k$. The system converges to the face containing equilibrium P_1 (see Table 14.1).

(ii) The new mutant, v_2, may not elicit an immune response against itself, but may induce a partial shift in immunodominance. This happens if $1/c_1 + 1/c_2 > 1/k$ and $r_1 > r_2$. The system converges to the face containing equilibrium P_4.

Table 14.1 *Saturated fixed points for the 2×1 system with $\eta = 0$ as specified by eqn (14.29)*

Fixed point	(v_1, v_2)	(x_1, x_2)	y	Conditions of existence and stability	
P_1	$(+, +)$	$(+, +)$	0	$1/c_1 + 1/c_2 < 1/k$	
P_2	$(+, +)$	$(0, +)$	+	$1/c_2 < 1/k < 1/c_1 + 1/c_2$	$r_1 < r_2$
P_3	$(0, +)$	$(0, 0)$	+	$1/k < 1/c_2$	$r_1 < r_2$
P_4	$(+, +)$	$(+, 0)$	+	$1/c_1 < 1/k < 1/c_1 + 1/c_2$	$r_1 > r_2$
P_5	$(+, 0)$	$(0, 0)$	+	$1/k < 1/c_1$	$r_1 > r_2$

The table shows the conditions of existence and stability of the five different stable equilibria, which occur for five distinct parameter regions. The sign of the coordinates of the fixed points are shown; the actual values can easily be calculated from eqn (14.29). The equilibria P_1, P_3, and P_5 specify complete immunodominance, whereas P_2 and P_4 imply stable coexistence between one of the x-responses and the y-response.

(iii) The new mutant may induce a response against itself, which outcompetes the response against the original virus, and induces a partial shift in immunodominance. The conditions for this behaviour are $1/c_1 + 1/c_2 > 1/k > 1/c_2$ and $r_1 < r_2$. We end up in the face containing equilibrium P_2.

(iv) Finally, the new mutant can induce a complete shift in immunodominance to epitope B. This happens for $1/c_2 > 1/k$ and $r_1 < r_2$, which brings us to the face containing equilibrium P_3.

Note that equilibrium P_5 is excluded by our original assumption that $c_1 > k$. In other words, a mutant in the immunodominant epitope can always invade.

In all four cases there will be undamped oscillations around the relevant equilibrium. Only the time averages will converge towards the equilibrium. In general these oscillations will be very complex. Figure 14.7 shows a computer simulation of eqn (14.29). The parameters are chosen such that there is convergence towards the face $x_1 = 0$. The system is specified by quasiperiodic oscillations. It is interesting to note that such complex, and unpredictable dynamics can occur for a system with only two virus variants and two immune responses against different epitopes.

14.5.2 $\eta > 0$

Next we consider the 2 × 1 system with positive η. For small η it is clear that there can be only one fixed point in the interior, because for $\eta > 0$ only the saturated fixed points of the system with $\eta = 0$ can migrate into the interior. Since the $\eta = 0$ system has only one saturated fixed point for any one choice of parameters, we can conclude that there can be only one saturated (and hence at most one interior) fixed point for the system with positive (but small) η. We have

$$
\begin{aligned}
\dot{v}_1 &= v_1(r_1 - p_1 x_1 - qy), \\
\dot{v}_2 &= v_2(r_2 - p_2 x_2 - qy), \\
\dot{x}_1 &= x_1(c_1 v_1 - b) + \eta c_1 v_1, \\
\dot{x}_2 &= x_2(c_2 v_2 - b) + \eta c_2 v_2, \\
\dot{y} &= y[k(v_1 + v_2) - b] + \eta k(v_1 + v_2).
\end{aligned}
\tag{14.30}
$$

If v_1 and v_2 are both strictly positive, there is no equilibrium without x_1, x_2 and y all being present. An interior equilibrium satisfies

$$
\begin{aligned}
x_i &= \frac{\eta c_i v_i}{b - c_i v_i}, \\
y &= \frac{\eta k(v_1 + v_2)}{b - k(v_1 + v_2)},
\end{aligned}
\tag{14.31}
$$

and

$$
r_i = p_i x_i + qy.
\tag{14.32}
$$

For this, we must of course have $v_i < b/c_i$ and $v_1 + v_2 < b/k$.

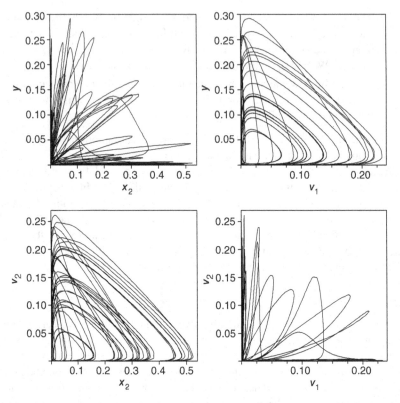

Fig. 14.7 The quasiperiodic behaviour of eqn (14.29), i.e. the 2×1 model with $\eta = 0$. The parameters are: $r_1 = 0.02$, $r_2 = 0.05$, $c_2 = 2$, $k = 1$ and $b = 0.02$. The parameter c_1 is chosen such that $1/k < 1/c_1 + 1/c_2$. This implies convergence to the face with $x_1 = 0$, which corresponds to case P_2 in Table 14.1. We are left with a subsystem containing the variables v_1, v_2, x_2 and y. This is the actual system we consider for the computer simulation. There is an invariant of motion (see Appendix B.1) which reduces the dimension to three. There we find quasiperiodic behaviour. The figure shows the time trajectories of y versus x_2, y versus v_1, v_2 versus x_2, and v_2 versus v_1.

There are two equilibria on the boundary of the state space. Let us denote by P_1 the equilibrium with $v_2 = x_2 = 0$ and by P_2 the equilibrium with $v_1 = x_1 = 0$. Let us consider P_1. The equilibrium must satisfy

$$r_1 = p_1 x_1 + qy, \quad x_1 = \frac{\eta c_1 v_1}{b - c_1 v_1}, \quad y = \frac{\eta k v_1}{b - k v_1} \quad (14.33)$$

with $v_1 < b/c_1$ and $v_1 < b/k$. This implies

$$r_1(b - c_1 v_1)(b - k v_1) = \eta p_1 c_1 v_1 (b - k v_1) + \eta q k v_1 (b - c_1 v_1), \quad (14.34)$$

i.e., after setting $v_1 = bw$ and dividing by $b^2 k c_1$,

$$f(w) \equiv r_1 \left(w - \frac{1}{k} \right) \left(w - \frac{1}{c_1} \right)$$

$$+ \eta w \left[p \left(w - \frac{1}{k} \right) + q \left(w - \frac{1}{c_1} \right) \right] = 0. \qquad (14.35)$$

Since f is negative at the smaller of the two values $1/c_1$ and $1/k$, and positive at the larger of the two values as well as at the origin, we have exactly one root between 0 and $\min\{1/c_1, 1/k\}$. This yields the desired equilibrium P_1 on the boundary face $x_2 = v_2 = 0$. A straightforward application of the Routh–Hurwitz criterion shows that P_1 is stable *within* the corresponding boundary face. It is saturated iff v_2 cannot invade, i.e. iff

$$r_2 < q \hat{y}_1. \qquad (14.36)$$

We note that it is impossible to have both P_1 and P_2 saturated since $q \hat{y}_1 < r_1$ and hence $r_2 < r_1$ whenever P_1 is saturated. Therefore, we cannot have a bistable situation with both boundary fixed points being saturated.

The condition (14.36) for saturation of the boundary fixed point P_1 can explicitly be written as:

(i) $r_1 > r_2$ and
(ii) if $c_1 > k$ then

$$\eta k q r_1 / r_2 > (r_1 - r_2)(c_1 - k) + \eta(kq + c_1 p_1) > r_2(c_1 - k), \qquad (14.37)$$

if $c_1 < k$ then

$$(r_1 - r_2)\left(k - c_1 + \frac{\eta k q}{r_2} \right) > \eta c_1 p_1. \qquad (14.38)$$

A similar condition determines saturation of P_2.

But even for the simple 2×1 system the general model with arbitrary η leaves some open questions. For example, we do not know if a saturated boundary equilibrium is compatible with the existence of an inner equilibrium (for the same choice of parameters). We also do not know if it is possible to have several interior equilibria. The answers to these questions seem to require explicit solutions of third- and fourth-order polynomials. However, we conjecture that there is always only one stable equilibrium. This is certainly true in the limit of small η, and in Section 14.5.3 we will show that it also holds in the limit of large η.

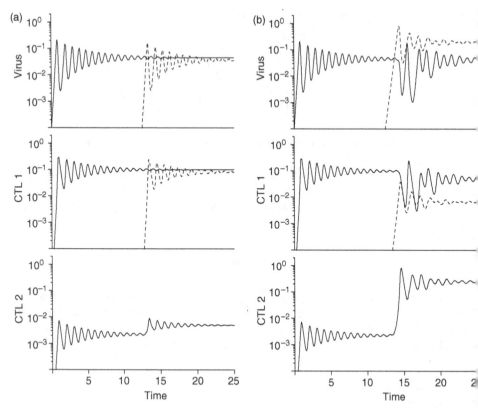

Fig. 14.8 The figure shows the four different dynamical possibilities following the emergence of a new mutant. Equation (14.30) with $\eta = 0.01$ is used for the computer simulations. Originally only one virus variant (v_1) is present. The x_1-response is immunodominant, because we chose $c_1 > k$. After some time mutant v_2 is introduced. There are four different possibilities which correspond to the four cases which were discussed for the $\eta = 0$ model. (a) The mutant induces an x_2-response but does not significantly affect the y-response. This is simply a diversification in epitope A, without changing immunodominance. (b) The mutant only induces a very weak x_2-response (maybe below an experimental detection limit), but stimulates a very strong y-response. The x_1-response is more or less unaffected. This corresponds to a partial shift in immunodominance.

Figure 14.8 shows computer simulations of system (14.30) for $\eta = 0.01$. All simulations are originally started with a homogeneous virus population (variant v_1). It is assumed that $c_1 > k$ such that the x_1-response is immunodominant. Since $\eta > 0$ the y-response survives, too, but at much lower levels. After some time mutant v_2 introduced. The figure shows four dynamical possibilities, which correspond to the four

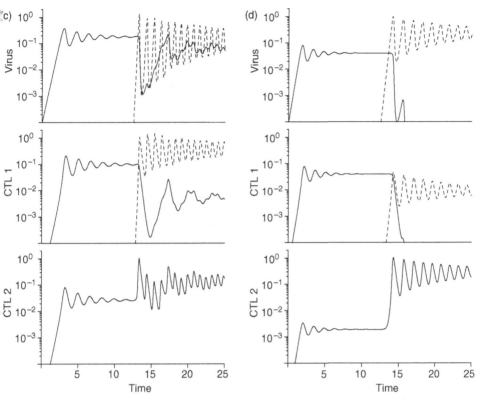

Fig. 14.8 (c) The mutant induces a strong x_2-response, reduces the x_1-response, and increases the y-response. (d) The mutant outcompetes the original v_1 variant (together with the x_1-response), induces only a very weak x_2-response, but a very strong y-response. This represents an almost complete shift of immunodominance. The parameters are: (a) $r_1 = 0.1$, $r_2 = 0.1$, $c_1 = 1$, $c_2 = 1.2$; (b) $r_1 = 0.1$, $r_2 = 0.05$, $c_1 = 1$, $c_2 = 0.1$; (c) $r_1 = 0.03$, $r_2 = 0.15$, $c_1 = 0.25$, $c_2 = 0.3$; (d) $r_1 = 0.04$, $r_2 = 0.06$, $c_1 = 1$, $c_2 = 0.1$; and $k = 0.2$, $b = 0.05$, $p_i = c_i$, $q = k$.

cases we have described analytically for $\eta = 0$. In Fig. 14.8(a) the new virus mutant induces a significant x_2-response, but does not affect the y-response. In Fig. 14.8(b) the new mutant induces only a very weak x_2-response (maybe below the limit of detection), but greatly enhances the y-response, thereby shifting immunodominance. In Fig. 14.8(c) the new mutant induces a strong x_2-response. The x_1-response is significantly weakened, but the y-response is somewhat enhanced. In Fig. 14.8(d) the new mutant outcompetes the original v_1 variant, induces only a weak x_2-response, but a strong y-response. This corresponds to a more or less complete shift in immunodominance.

14.5.3   The limit of large η

In the limit of large η we can write (see Section 14.4.3)

$$
\begin{aligned}
\dot{v}_1 &= v_1(r_1 - p_1 x_1 - qy), \\
\dot{v}_2 &= v_2(r_2 - p_2 x_2 - qy), \\
\dot{x}_1 &= \eta c_1 v_1 - b x_1, \\
\dot{x}_2 &= \eta c_2 v_2 - b x_2, \\
\dot{y} &= \eta k(v_1 + v_2) - by.
\end{aligned}
\tag{14.39}
$$

This system has three non-trivial equilibria: one fixed point in the interior and two boundary fixed points. There are three parameter regions, which are mutually exclusive. For each parameter region exactly one of these fixed points is stable:

- If $r_1/r_2 > (c_1 p_1 + kq)/kq$ then v_2 and x_2 converge to zero.
- If $kq/(c_2 p_2 + kq) > r_1/r_2$ then v_1 and x_1 converge to zero.
- If $(c_1 p_1 + kq)/kq > r_1/r_2 > kq/(c_2 p_2 + kq)$ then the interior fixed point is stable.

Again for any choice of parameters there is exactly one stable equilibrium.

In Appendix B.3 we discuss the 2×2 system and give a complete classification for the two limiting cases $\eta = 0$ and η very large.

14.6   Cross-reactivity within the variants of a given epitope

In this section we analyse the effect of cross-reactivity within the sequences of a given epitope. We assume that sequence i of epitope A can cross-stimulate the response against sequence j of epitope A, at a rate c_{ij}. Similarly we define k_{ij} for all the variants of epitope B. We include this cross-reactivity in both the stimulation term and in the term for the CTL response against the virus. This leads to

$$
\begin{aligned}
\dot{v}_{ij} &= v_{ij}\left(r - p \sum_{l=1}^{n_1} c_{il} x_l - q \sum_{l=1}^{n_2} k_{jl} y_l \right), \\
\dot{x}_i &= x_i\left(\sum_{l=1}^{n_1} c_{il} v_{l*} - b \right) \quad \text{with } i = 1, \ldots, n_1, \\
\dot{y}_j &= y_j\left(\sum_{l=1}^{n_2} k_{jl} v_{*l} - b \right) \quad \text{with } j = 1, \ldots, n_2.
\end{aligned}
\tag{14.40}
$$

The outcome depends on the $n_1 \times n_1$ matrix $\{c_{ij}\}$ and the $n_2 \times n_2$ matrix $\{k_{ij}\}$; the earlier simpler model is recovered as the limit $c_{ij} \to c_i \delta_{ij}$, etc. As an example, let us assume a very simple form for these matrices: $c_{ii} = c$, $c_{ij} = cs_1$, $k_{jj} = k$ and $k_{ij} = ks_2$ for all values of i and j with $i \neq j$. The parameters c and k denote the immunogenicities of epitopes A and B as in the previous sections, and the parameters s_1 and s_2 specify the amount of cross-reactivity within epitopes A and B, respectively. If $s_1 = 1$ there is complete cross-reactivity, and if $s_1 = 0$ there is no cross-reactivity (for epitope A). By generalizing the arguments which lead to eqn (14.6), we find that the response against A will eventually dominate over the response against B if

$$\frac{c}{n_1}[1 + (n_1 - 1)s_1] > \frac{k}{n_2}[1 + (n_2 - 1)s_2]. \tag{14.41}$$

This result shows that for the initial phase of an infection, when the diversity is low (i.e. n_1 and n_2 around one), the decisive parameter is immunogenicity (c versus k). At later stages of infection, when both n_1 and n_2 have increased, cross-reactivity becomes important. This means that during an HIV infection there may be a tendency to go from responses against highly immunogenic epitopes towards (possibly) less immunogenic, but more cross-reactive, epitopes. Here cross-reactivity is always defined within the various sequences of a given epitope. We do not consider cross-reactivity among different epitopes. These more general effects of cross-reactivity are, to some degree, captured by the cross-reactive immune response in the previous chapters (see Chapters 12 and 13).

14.7 Immunogenicity and intracellular competition

The immunogenicity of an epitope may be affected by mutations that occur in other epitopes. We can easily imagine a mutation that enhances a peptide's affinity for MHC binding. If there is some intracellular competition for MHC binding, then this could reduce the overall MHC presentation of another epitope. Thus in more general terms the immunogenicity of an epitope is not only a function of the particular peptide sequence, but also of the sequences of other epitopes (or the protein or pathogen as a whole). A model that takes this into account has the following form:

$$\dot{v}_{ij} = v_{ij}(r - x_i - y_j),$$
$$\dot{x}_i = x_i\left(\sum_{l=1}^{n_2} c_{il}v_{il} - b\right), \tag{14.42}$$
$$\dot{y}_j = y_j\left(\sum_{l=1}^{n_1} k_{lj}v_{lj} - b\right).$$

(For simplicity we present the case $\eta = 0$. Including a positive η-term adds the usual complications.) We also restrict ourselves to the case where all virus mutants have the same replication rate, r. The $n_1 \times n_2$ matrix $\{c_{ij}\}$ has elements which denote the immunogenicity of sequence i in epitope A of the virus v_{ij} (with sequence j in epitope B). Similarly k_{ij} is the immunogenicity of sequence j in epitope B given that there is sequence i in epitope A. The earlier, simpler model is recovered as the limit $c_{ij} \rightarrow c_i$ and $k_{ij} \rightarrow k_j$.

In general, eqn (14.42) admits three sets of equilibria: one set of interior equilibria with all x_i and y_j being positive and two sets of boundary equilibria with either all x_i or all y_j being zero. For the same argument as in Section 14.3, eqn (14.9), it is not possible to have a stable equilibrium with some y_j being zero and others positive. But contrary to system (14.4), interior equilibria with both all y_j and all x_i responses positive can exist. The three sets of equilibria are given by the relations:

1. Interior equilibria:

$$x_i = \xi, \quad y_j = r - \xi, \quad \sum_{j=1}^{n_2} c_{ij} v_{ij} = b, \quad \sum_{i=1}^{n_1} k_{ij} v_{ij} = b, \quad \text{for all } i, j. \quad (14.43)$$

Here ξ is some arbitrary number between 0 and r. There are $n_1 + n_2$ equations for the $n_1 \times n_2$ variables, v_{ij}. Thus in general such interior equilibria may exist. Our numerical simulations suggest that these equilibria are not stable, with the system exhibiting either collapse to a boundary equilibrium (see below) or heteroclinic cycles. In Appendix B.4 we give analytic results, showing there cannot be convergence to an interior equilibrium; at best interior states can have locally neutral stability.

2. Boundary equilibria with all $y_j = 0$:

$$y_j = 0, \quad \text{for all } j; \quad x_i = r, \quad \sum_{j=1}^{n_2} c_{ij} v_{ij} = b, \quad \text{for all } i. \quad (14.44)$$

In the face with all $y_j = 0$ these equilibria are surrounded by neutral oscillations. There is saturation with respect to invasion by the y_j if

$$\sum_{i=1}^{n_1} k_{ij} v_{ij} < b, \quad \text{for all } j. \quad (14.45)$$

The unsolved problem is that some of the equilibria (depending on the particular v_{ij}-configuration) may be saturated while some may be unsaturated. Numerical simulations (and analytic investigations of special cases; see below) suggest that it depends on the initial conditions for the v_{ij}, whether or not there is convergence to this subset of saturated equilibria (with all $y_j = 0$).

3. Boundary equilibria with all $x_i = 0$:

$$x_i = 0, \quad \text{for all } i; \quad y_j = r, \quad \sum_{i=1}^{n_1} k_{ij} v_{ij} = b, \quad \text{for all } j. \quad (14.46)$$

Saturation against invasion by the x_i requires

$$\sum_{i=1}^{n_1} c_{ij} v_{ij} < b, \quad \text{for all } i. \quad (14.47)$$

Again some equilibria may be saturated and some may be unsaturated. Convergence appears to depend on the initial conditions.

We can derive a complete analytical understanding for a simple case with symmetric initial conditions. Let us consider eqn (14.42) with $n_1 = n_2 = 2, c_{ij} = 1, k_{11} = k_{22} = k$, and $k_{21} = k_{12} = \kappa$. Consider the initial conditions $v_{11}(0) = v_{22}(0)$ and $v_{12}(0) = v_{21}(0)$. Let us define $\beta \equiv v_{12}(0)/v_{11}(0)$. Because of symmetry we have $v_{11}(t) = v_{22}(t)$, $v_{12}(t) = v_{21}(t)$, $x_1(t) = x_2(t)$, $y_1(t) = y_2(t)$ and $v_{12}(t)/v_{11}(t) = \beta$. Thus the system collapses to only three independent dimensions:

$$\dot{v} = v(1 - x - y),$$
$$\dot{x} = x[(1 + \beta)v - b], \qquad (14.48)$$
$$\dot{y} = y[(k + \beta\kappa)v - b].$$

We have defined $v \equiv v_{11}, x \equiv x_1, y \equiv y_1$. Equation (14.48) has the solution

$$x^{k+\beta\kappa} y^{-(1+\beta)} = C \exp[(1 - k + \beta(1 - \kappa))bt]. \qquad (14.49)$$

Here C is a constant specified by the initial conditions. Thus for $t \to \infty$, the x-responses die out if $1 - k + \beta(1 - \kappa) < 0$; while the y-responses die out if the opposite inequality holds. Thus it depends on the initial ratio, $\beta = v_{12}(0)/v_{11}(0)$, whether the x- or y-responses become immunodominant. Coexistence occurs only for a set of initial conditions with measure 0, namely if $1 - k + \beta(1 - \kappa) = 0$.

The important point of this section is that in situations where the immunogenicity of an epitope can be affected by changes in other epitopes, immunodominance may depend on the initial configuration of the virus population and is thereby largely determined by chance events.

14.8   Immunotherapy

Several interesting hints emerge from the above analyses relevant to the design of a potential post-exposure vaccine. In general, immunotherapy should be directed at conserved epitopes. Remember that it is only necessary to control the virus population in a single epitope. But let us assume that there is a conflict in the sense that the highly immunogenic epitopes display antigenic variation, but the conserved epitopes are only weakly immunogenic. If the response against the weakly immunogenic but conserved epitope can be enhanced, such that this epitope becomes immunodominant, then the virus population will be controlled by the response against this epitope. Variation that may occur in other epitopes can then only reinforce the immunodominance of this conserved epitope. If immunotherapy is not potent enough to achieve immunodominance of the response against the conserved epitope, then it may be advantageous to direct immunological attack at the more immunogenic but variable epitopes. Let us suppose that the immunogen will only stimulate responses against a certain fraction of the occurring variants. If these variants differ in their intrinsic immunogenicities, then it is always better to enhance the responses against the weakly immunogenic variants.

These points are illustrated explicitly by the following equations. Let us again consider the 2×1 case as given by eqn (14.29). Now suppose that immunotherapy against

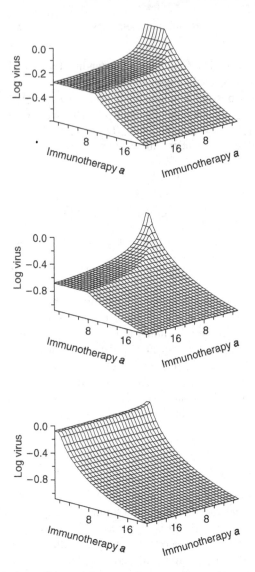

Fig. 14.9 The relative effect of immunotherapy directed at the variable epitope A, or the conserved epitope B. As model we use the simple 2×1 case with $\eta = 0$ as described by eqn (14.29). Immunotherapy against A can only recognize sequence 1 and enhances its immunogenicity by a factor α. Immunotherapy against B enhances the immunogenicity of this peptide by a factor β. The figure shows the equilibrium virus load as a function of the efficacies of immunotherapy, α and β. If $1/(\alpha c_1) + 1/c_2 < 1/\beta k$ then the equilibrium virus load is $v = b[1/(\alpha c_1) + 1/c_2]$. If $1/(\alpha c_1) + 1/c_2 > 1/\beta k$ then the equilibrium virus load is $v = b/(\beta k)$. The basic message is that an effective control with immunotherapy can only be achieved by stimulating responses against conserved epitopes. If it is not possible to make the conserved epitopes immunodominant, than immunotherapy against the variable epitopes may be preferable, but here immunotherapy should always be directed

the variable epitope A can enhance the immunogenicity of variant 1 from c_1 to αc_1, with $\alpha > 1$. Variant 2 is not recognized. Immunotherapy against the conserved epitope B enhances its immunogenicity from k to βk, with $\beta > 1$. If $1/(\alpha c_1) + 1/c_2 < 1/(\beta k)$ then the equilibrium virus load is $v = b[1/(\alpha c_1) + 1/c_2]$. If conversely $1/(\alpha c_1) + 1/c_2 > 1/(\beta k)$ then the equilibrium virus load is $v = b/(\beta k)$. Figure 14.9 gives an illustration of the equilibrium virus load as a function of the efficacies of the immunotherapies against the two epitopes. For high efficacy the virus population levels off if immunotherapy is directed at the variable epitope A, while it would still decline if the response is directed at B. For very low efficacy it may sometimes be advantageous to induce responses against the more immunogenic, but variable, epitope A.

14.9 Summary

In this chapter we developed a mathematical theory for immune responses against multiple epitopes. The theory has originally been designed to explain the dynamical interaction between CTL responses and the variable HIV quasispecies, but has a wider potential. It represents a mathematical framework for any kind of specific immune response (CD8+, CD4+, or antibody responses) against multiple epitopes of a replicating and variable pathogen. The principal conclusions are as follows:

1. Antigenic oscillations can arise as a consequence of the dynamics of the immune response acting upon existing viral diversity. It is not essential that mutation continuously generates new antigenic material. Peaks consisting predominantly of different antigenic types can rise and fall as a consequence of the oscillatory dynamics; whenever the CTL response against a particular variant has fallen to lower levels, this variant may start to grow and cause a new peak.

2. Immunodominance is a function both of the immunogenicity and of the antigenic diversity of the epitopes, and also of the replication rate of the various mutants. If there is only a homogeneous virus population, then the generic situation is that there is one immunodominant epitope. For an antigenically heterogeneous virus population we may find coexisting responses against several epitopes. But if there is a heterogeneous virus population and all virus mutants have the same replication rate, then again generically there is always a single immunodominant epitope. Only the response against one epitope can survive; all other responses have to vanish. The epitope that minimizes $\sum_{i=1}^{n} 1/c_i$ is immunodominant (with c_i being the immunogenicity of sequence i, and n being the total number of variants in this epitope). If the virus mutants have different replication rates, then the conditions for immunodominance are more complex (see Section 14.3.1). If all responses are directed against a single epitope, then this must be the epitope which minimizes $\sum 1/c_i$. But this condition, although necessary, is not sufficient. A number

---

at the weakly immunogenic sequences, too (compare (b) to (c)). Parameters: (a) $c_1 = 10$, $c_2 = 10$; (a) $c_1 = 2$, $c_2 = 10$; (a) $c_1 = 10$, $c_2 = 2$; $k = 1$. The replication rates are irrelevant.

of inequalities have to be fulfilled, and the replication rates of the individual mutants become important. As a rule of thumb, we expect coexistence of immune responses against several epitopes, if the immunogenicities of the various epitopes are comparable and if the virus mutants differ in their overall replication rates.

In general this picture is also supported by observations of HTLV-I infections, where responses against several epitopes coexist, stimulated by an antigenically diverse virus population (see Parker *et al.* 1994).

3. Antigenic variation (i.e. production of new antigenic material) can shift immunodominance. The emergence of a very weakly recognized sequence in an epitope does not necessarily lead to this sequence dominating the population, but will lead to a shift in immunodominance to another epitope. This possibly explains why it has been occasionally observed in HIV infection that CTL escape mutants do not grow to dominate the population.

4. One of the central points to emerge from this chapter is a clear understanding of the events following the emergence of a new mutant. Consider a homogeneous virus population subject to immune responses against two epitopes, A and B. Suppose that the response against A is immunodominant. The emergence of an escape mutant in A can lead one of four different outcomes, depending on the relative replication rates and immunogenicities of the original virus and the new mutant: (i) the new mutant may induce a new specific response in epitope A, without affecting the response against B (this represents a simple diversification in epitope A); (ii) the new mutant may not induce a response in A against itself, but may enhance the response against epitope B (this corresponds to a partial shift in immunodominance); (iii) the new mutant may induce a response in A against itself, which outcompetes the original response in A (this always occurs together with an increase of the response against epitope B, thus representing a partial shift in immunodominance); (iv) finally, the new mutant may outcompete the original virus variant, and induce a complete shift in immunodominance to epitope B (the response against A essentially vanishes). This has important consequences for our understanding of the detailed escape dynamics in the presence of responses against multiple epitopes.

5. Shifting immunodominance to intrinsically weaker epitopes increases viral loads, and can thus represent a route to disease progression in HIV infection.

6. Clearly the models presented here are not limited to HIV, nor to any particular virus. Any (fast) replicating (variable) pathogen with several epitopes is a relevant target for our mathematical framework. Escape from CTL recognition has been demonstrated in a number of human virus infections, such as HIV-1 (Phillips *et al.* 1991), HTLV-1 (Parker *et al.* 1994), hepatitis B virus (Bertoletti *et al.* 1994) and Epstein–Barr virus (Campos-Lima *et al.* 1993). Detailed *in vivo* and *in vitro* studies also exist for lymphocytic choriomeningitis virus (Aebischer *et al.* 1991). Antigenic diversity in CTL epitopes has also been found in human malaria (Hill *et al.* 1992). Our basic model has outlined the competitive dynamics of simultaneous immune (CTL) responses against multiple epitopes; it gives a quantitative concept of immunodominance. Furthermore, the model is not limited to CTL responses. Antibody or CD4+ T-helper responses are likely to

obey the same underlying mathematical rules. We have concentrated on CTL responses simply because we think that here the biology is best understood.

7. With respect to HIV, our models reinforce the notion that viral diversity is important for pathogenesis. There are obvious effects of antigenic diversity on viral levels, and hence on disease progression, even without invoking the viral-induced depletion of CD4 cells (which is essential for the diversity threshold theory). We have shown that antigenic diversification can shift immunological pressure towards weaker epitopes and therefore increase virus load.

14.10   Further reading

Multiple epitope theory is described by Nowak *et al.* (1995*a,b*), Nowak (1996), Nowak and McMichael (1995). For further experimental evidence of antigenic variation during HIV or other virus infections refer to Aebischer *et al.* (1991), Bertoletti *et al.* (1994), Borrow *et al.* (1997*a,b*), Campos-Lima *et al.* (1993), Carpenter *et al.* (1990), Domingo *et al.* (1993), Goulder *et al.* (1997), Klenerman *et al.* (1994), McMichael (1993), Nara *et al.* (1990), Parker *et al.* (1994), Phillips *et al.* (1991), Price *et al.* (1997), Salinovich *et al.* (1986), and Wahlberg *et al.* (1991). See Sasaki (1994) for an excellent mathematical model of antigenic drift. For experimental work on immunodominance see Adorini *et al.* (1988). Gupta *et al.* (1996, 1998) and Gupta and Anderson (1999) describe multiple epitopes and antigenic variation in epidemiological dynamics.

15

EVERYTHING WE KNOW SO FAR AND BEYOND

In this final chapter, we provide a summary of our results and outline some open questions. The general purpose of this book was to provide a mathematical framework for how viruses interact with the immune system. We developed a theory for the population dynamics or *micro-epidemiology* of infectious agents within infected individuals.

15.1 The mechanism of HIV-1 disease progression

Large parts of the book were devoted toward describing the dynamics of HIV infection. In this section, we review the question to what extent we have an understanding of HIV disease progression.

As we have seen, HIV infects and kills CD4 positive T-helper cells. Soon after infection, patients undergo a *primary phase*: at first the virus grows exponentially, reaches a peak and then starts to decline. Sometime during the primary phase of infection, patients mount CTL and antibody immune responses against the virus. It is believed that these immune responses are responsible for the observed decline in virus load, but we know that simple virus dynamics equations without immune responses can also display oscillatory behaviour and in particular give rise to an initial peak in virus load followed by a decline to a steady-state. Thus it is unclear to what extent the immune responses are involved in the reduction of virus load during primary HIV-1 infection.

During the *asymptomatic phase* of infection, there is ongoing virus replication. Productively infected cells live on average for about 2 days. These cells generate roughly 99% of the virions present in the circulation of a patient. In addition to productively infected cells, there are latently infected cells and cells harbouring defective HIV genomes. Latently infected cells have half-lives of about 10–40 days, while cells with defective HIV provirus live for about 100 days. There may be different kinds of 'latently infected cell': those that are truly latent and waiting to be reactivated and those that are slow-chronic producers of virions. (The kinetics of HIV infection in macrophages is poorly understood.) Free virions seem to be rapidly eliminated from the plasma of infected patients (or experimetally infected animals): their half-life seems to be less than a few hours. Thus most of the virions that can be found in the plasma of a patient were generated during the last few hours; and 30% of those cells that produced them are replenished every day.

These figures represent some quantitative insights into viral kinetics, but the overall picture remains incomplete. Most importantly, no one so far has been able to quantify the effect of immune responses on the observed dynamics. Are productively infected cells killed after 2 days by virus cytopathicity or CTL-mediated lysis? Are CTL involved in the removal of latently infected cells? Do antibody mediated immune responses contribute to the clearance of plasma virions? Such questions have a common theme: they demand a quantitative understanding of immune responses in HIV infection. Developing experimental techniques and theoretical concepts for answering these questions, seems to be the most important next step.

The quantitative insights into viral demography have important consequences for understanding viral turnover, anti-viral treatment and the kinetics of drug-resistance or escape from immune responses, but they do not resolve the question of HIV disease progression. More precisely, the measured virus dynamics tell us something about the nature of the steady-state between the virus and the immune system, but do not tell us anything about what shifts this steady-state, slowly over many years, in favour of the virus. Understanding disease progression in HIV infection means to know the mechanism that shifts the steady-state. No one, so far, understands HIV disease progression.

What we set out to do, about 10 years ago, in 1989, was to develop a mathematical framework for the *in vivo* population dynamics of infectious agents and the immune system and specifically to study HIV infection. Judging from observations of primary infection, AZT treatment and the rapid genetic variation of HIV-1 *in vivo*, we assumed that viral turnover had to be fast, on the time scale of days. Thus, there would be a steady-state between the virus and the immune system, and the main question was what changes this dynamic balance over the time of years. We suggested that this mechanism was virus evolution that occurs relentlessly during individual infections.

From what we now know about HIV infection, such an evolutionary mechanism of disease progression is a reasonable possibility. The large virus population size in an infected patient and the short generation time (of about 2 days) makes HIV extremely flexible for evolutionary adaptation. In general, there will be competition among different HIV mutants for increasing reproduction rates. The virus will adapt to various ecological niches within the body, and 'learn' to infect a number of different cell types. This process will broaden the cell tropism of HIV and will during the time of infection generate faster growing, more virulent variants. We expect, however, that this process is not independent of the immunological pressure exerted on the virus. It is likely that a strong immune response against the virus selects for mutants that can efficiently avoid this immune response, while in a patient with a weak immune response, there will be simple selection for fast replication.

Furthermore, there is convincing evidence that HIV generates mutants that escape from specific antibody or T-cell responses. We have in some detail explained how such a process can give rise to disease progression. In a multiple epitope setting, antigenic variation in an immunodominant epitope can divert immune responses to other, weaker epitopes. This reduces the contol the immune system has over the virus and leads to higher virus load. In general, antigenic variation will work over time to increase the virus load in a patient, thereby shifting the dynamic balance in favour of the virus. Antigenic variation during individual infections is certainly not only a property of HIV, but is a feature of many fast replicating, flexible viruses or other infectious agents. Therefore, the concept that antigenic variation enhances virus load should apply to all such infections.

In addition, however, the mathematical models for HIV suggest that there is an antigenic diversity threshold: if the antigenic diversity of the virus population in a patient exceeds a certain value then the immune system fails to control the virus population. This threshold arises for any pathogen that can impair the immune system (in addition to producing antigenic variants, of course). The threshold is not a trivial statement in the sense that the diversity of the virus population can outrun the diversity of the immune system. Instead it is a property of the asymmetric interaction between HIV and the immune system: specific immune responses are directed against specific virus mutants, but all virus mutants can impair all immune responses.

Patients who mount weak immune responses against HIV have low thresholds. They might progress to disease rapidly and without much antigenic variation. In contrast, patients with strong immune responses have high thresholds and will progress slowly and with substantial antigenic variation. This complication should not be overlooked when testing the evolutionary theory of HIV disease progression.

It is, of course, possible that the evolutionary model fails to provide an adequate mechanism for HIV disease progression. This would be the case if the immune response would not at all control the virus or only mount responses which act in a very cross-reactive way against all mutants that are present. In this case, an evolutionary model of disease progression could only be based on selection for faster replication rates and broader cell tropism and might be quite different to what we envisage. There is, however, no evidence that this is the case.

What are other possible mechanisms of HIV disease progression? McLean and Nowak (1992b) formulated a model according to which the steady-state between HIV and the immune system is shifted over time by increasing antigenic activation of CD4 cells. This immune activation could be a consequence of an accumulation of other pathogens that fail to be cleared by an immune system that is weakened by HIV. As more and more CD4 cells become activated, the virus population will find more target cells that support productive replication and hence reproduce faster and grow to higher levels of virus load. In such a description, it is not necessary for the virus quasispecies to change over time in any important way, but it is easy to see how the immune activation model could be combined with an evolutionary model. The two models are not mutually exclusive.

A third class of mechanisms can be formulated as stating that the explanation is found completely within the immune system. If a virus like HIV kills a certain number of CD4 cells per day then the immune system will break down after some time. Such a statement, however, is simply a reformulation of the observation that CD4 cell numbers decline slowly over time. Hence, such a mechanism is somewhat of a tautology and only becomes a reasonable hypothesis once we have any idea why the immune system should succumb so slowly to HIV infection. In other words, the big open question of such a mechanism is why can the immune system replace almost all, but not all of the CD4 cells that are killed by HIV per day and why is the difference so small? In an average HIV infection, over a long time span about 0.3 CD4 cells per ml blood (or 0.03% of the CD4 cell count of a healthy person) are lost per day.

Following a suggestion by John Leonard, Arnaout et al. (1999) looked at data of HIV cohort studies with a very simple-minded hypothesis: suppose the overall pattern of HIV disease progression is very simple, suppose CD4 numbers fall linearly over time while virus load is roughly constant during the asymptomatic phase of infection. If virus load is high then CD4 numbers decline rapidly, if virus load is low, then CD4 numbers decline slowly. The most elegant finding would be if the area under the virus load curve for all patients were constant. In other words, could it be that the total amount of virus experienced by each patient—either fast or slow progressor—was roughly the same? Ramy Arnaout and colleagues found some evidence into this direction, but their data set was not large enough to be certain about this finding. Larger cohorts of patients need to be analysed. In the meanwhile, the interesting open question is, suppose that the total amount of virus experienced by each patient is roughly constant ('the Leonard constant') or suppose there is an upper limit, what does this tell us about the mechanism of HIV disease progression? Does such a relationship favour any particular mechanism?

So far we have argued that understanding HIV pathogenesis means to understand the mechanism that shifts the dynamic balance between HIV and the immune system during the course of infection toward lower CD4 cell count and higher virus load. There is another way to look at this problem. Understanding the mechanism of pathogenesis should also enable us to explain the difference between fast and slow disease progression. Furthermore, we would like to understand the difference between the seemingly apathogenic, natural lentivirus infections and the pathogenic HIV infection of humans.

We can certainly list a number of viral and host factors that should influence the rate of disease progression in HIV infection. Certain virus strains might have the ability to replicate in a large number of cell types or to replicate rapidly in a particular cell type and thereby establish high virus load. Certain hosts may have mutations in viral receptor genes which reduce the virus' ability to grow well. The search for 'disease resistance genes' is on. There should also be particular, HLA-types which are associated with slow or fast disease progression. Charles Bangham's group in London found a clear cut example of a protective HLA allele in HTLV-1 infection. Steve O'Brien found deletions in chemokine receptor genes that seem to confer protection against infection by HIV-1. Jeff Lifson showed that the *in vitro* permisiveness of target cells prior to infection correlates with the rate of disease progression in SIV infection of macaques. There is also evidence that the amount of other infections carried by an HIV infected patient accelerates disease progression.

An interesting, new hypothesis was put forward by Wodarz *et al.* (2000). Slow progressors or long-term non-progressors might be patients that mount an effective CTL *memory* response against the virus. The new idea emerging from the mathematical models is that a long-lived CTL responses are required to clear viral infections, or failing this to reduce virus load to very low levels. The intuition behind this reasoning is as follows: a short-lived (non-memory) CTL response declines when antigen declines, a long-lived (or memory) CTL response remains active when antigen declines. A short-lived response leads to an equilibrium between the virus and the immune response, hence to a persistent virus infection, a long-lived response can clear the virus infection (or control it at very low levels). There is evidence that the generation of memory CTL needs CD4 cell help. Patients with impaired CD4-cell responses may still mount a CTL response, but it will not be of the memory phenotype. Thus HIV's ability to eliminate CD4 cells interfers with the development of a memory response. The consequence is a race between the virus and the immune system during primary infection. If the virus is faster then there will be no memory response resulting in high virus load and low disease progression. If the immune system is faster, there will be a memory response resulting in low virus load and slow disease progression. The important additional claim of the model is that treatment during primary infection, or intermittent treatment during chronic infection could switch the patient from one equilibrium to the other, that is from fast to slow disease progression.

15.2   How to overcome HIV

At the dawn of the new millenium, the HIV epidemic has reached roughly 50 million people. About 15 million have died and 35 million are currently infected. New infections arise at an enormous rate. Currently, our best weapons against HIV are drug combinations that inhibit virus growth in patients who are already infected. We have no vaccine against the virus, and although educational programs have had important effects among particular groups in particular places, by and large they had had little overall effect on slowing down the global spread of the HIV pandemic. While combination therapy offers great hope to HIV infected people in developed countries it will not have an impact in developing countries. Since most HIV infected people live in the world's poorest countries, the expensive combination therapy will not change the global picture. Thus the search for a simple and effective vaccine is still the most important goal of HIV research.

In the meantime, we have to learn how to make the best possible use of combination therapy. A major obstacle is the emergence of resistant virus. This process, however, is based on simple and precise principles: mutation and selection of viral quasispecies. Therefore, mathematical models can help to understand the emergence of resistance in great detail. We have argued that viral resistance is not a property of the virus alone, but a combination of viral and host factors. In precise terms, whether or not a specific virus mutant is resistant to a particular treatment regime is determined by the magnitude of the basic reproductive ratio of this mutant during treatment. If the basic reproductive ratio exceeds one, then the mutant can persist during treatment. In this case the virus population will remain at some equilibrium value and viral evolution during treatment will generate better adapted mutants that are more and more resistant during treatment. If, on the other hand, all virus mutants present in a patient at the time when treatment starts have a basic reproductive ratio below one, then the probability may be small that resistant mutants will emerge during therapy. Hence, the primary objective of anti-viral therapy should be to minimize the chance that resistant mutants are present at the initiation of therapy. This can be done by treating patients early in infection, when virus load is low and anti-viral immune responses are high, and using a potent combination of several drugs.

Given the long half-life of HIV provirus, which is a measure for the half-life of the longest-lived subpopulation of latently infected cells, eradication of chronic HIV infection by combination therapy alone does not seem to be a practical option. There are, however, two possibilities for a significant improvement of anti-HIV therapy. Combination therapy could be given in conjunction with drugs that activate latently infected cells and thereby accelerate the decline of this long-lived viral compartment. The main problem of this approach is that it might not be possible to stimulate all subpopulations of latently infected cells, and hence the virus could rise again once treatment is withdrawn. Another possibility is to use anti-viral treatment in order to stimulate a lasting memory CTL response against the virus. As mentioned above, this can be done by treating patients early during infection, by using intermittent therapy or by combining anti-viral treatment with immunotherapy. The hope of this approach is to switch patients into a state of long-term non-progression.

15.3   A quantitative immunology and virology

The last few decades have witnessed a tremendous advance of our understanding of the immune system. The elegant techniques of molecular biology have enabled immunologists to study the components of the vertebrate immune system in fascinating detail. Much is known about the molecular interactions among cells of the immune system and invading pathogens. Nevertheless, only very little is known about the dynamic interaction between populations of immune cells and populations of infectious agents. It is often possible to measure whether a particular immune response against, say, a viral epitope is present in a patient, but is almost never possible to be sure what effect it has. In other words, no one knows how many infected cells or viral particles are eliminated by a particular immune response over a certain time period. No one can quantify the *force of an immune response*, and yet it is clear that for a detailed mechanistic description of immunology this is the first question we have to ask.

Thus it seems to us that the major leap in immunology will come when it is possible to answer such questions. This will also enable us to move away from measuring correlates of immunity toward measuring actual immunity. The development of a vaccine will then become an engineering like approach. Clinical trials of drugs and vaccines, will then be supported by extensive and detailed computer simulations, so called 'virtual trials'.

We hope that the mathematical models outlined in this book provide the basic theoretical principles for such a quantitative approach to immunology.

15.4   Further reading

Reviews of HIV pathogenesis are Haase (1986), Coffin (1995), Haynes *et al.* (1996), Fauci (1996). Mathematical models of HIV pathogenesis are Nowak *et al.* (1990, 1991), Nowak and May (1993), McLean and Nowak (1992b), McLean (1993), Perelson *et al.* (1993), Levin and Bull (1994), Schenzle (1994), Mittler *et al.* (1996), Wodarz *et al.* (1998, 1999a), Callaway *et al.* (1999), Wodarz and Nowak (1999). Phillips (1996) describes a mathematical model for primary HIV infection. Mellors *et al.* (1996) and Arnaout *et al.* (1999) study the relation between HIV load and disease progression. Dean *et al.* (1996), Smith *et al.* (1997), Winkler *et al.* (1998), Martin *et al.* (1998), Carrington *et al.* (1999) describe mutations that affect HIV infection. Jeffery *et al.* (1999) describe protective HLA types in HTLV-1 infection. See Lifson *et al.* (2000) for containment of SIV infection.

APPENDIX A

DYNAMICS OF RESISTANCE IN DIFFERENT TYPES OF INFECTED CELLS

In this appendix, we discuss analytic approximations for the rise of resistant virus in different compartments of infected cells. We consider productively infected cells that produce large quantities of free virus in a short time, longer lived latently infected cells, and defectively infected cells that contain mutated virus genome and cannot produce new virions. This is, in essentials, further refinement of the basic models developed and discussed in Chapter 11.

Let y_1, y_2, and y_3 denote infected cells that contain active, latent or defective virus. We obtain

$$
\begin{aligned}
\dot{x} &= \lambda - dx - \beta xv, \\
\dot{y}_i &= q_i \beta xv - a_i y_i, \quad i = 1, 2, 3, \\
\dot{v} &= ky_1 + cy_2 - uv.
\end{aligned}
\tag{A.1}
$$

The parameter q_i describes the probability that upon infection a cell will enter type i; $\sum q_i = 1$. Thus q_1 is the probability that the cell will immediately enter active viral replication; y_1 cells will produce virus at rate k. The parameter q_2 is the probability that the cell will become latently infected with the virus and produce virus at a much slower rate c. In terms of this model, latent cells are in fact slow chronic producers of free virus. (More precisely, they should not produce any virus but turn back into productively infected cells after some time. Here we use a simplified approach to obtain analytic approximations.) The parameter q_3 specifies the probability that infection of a cell produces a defective provirus that will not produce any offspring virus. The decay rates of actively producing cells, chronically infected cells and defectively infected cells are a_1, a_2, and a_3, respectively. From Chapter 4, we know that a_1 is around 0.4 per day and a_3 around 0.01 per day. We expect a_2 to lie between these two values. The death rate of uninfected cells, d, will be similar to a_3 (probably slightly smaller).

Provided the basic reproductive ratio of the wild-type,

$$
R_0 = \frac{\lambda \beta A}{du},
\tag{A.2}
$$

is larger than one, the system converges to the equilibrium

$$
x^* = \frac{u}{\beta A},
$$

$$
y_1^* = \frac{q_1}{a_1}\left(\lambda - \frac{du}{\beta A}\right),
$$

$$y_2^* = \frac{a_1}{a_2}\frac{q_2}{q_1}y_1^*, \tag{A.3}$$

$$y_3^* = \frac{a_1}{a_3}\frac{q_3}{q_1}y_1^*,$$

$$v^* = \frac{\lambda}{u}A - \frac{d}{\beta},$$

where

$$A = \frac{kq_1}{a_1} + \frac{cq_2}{a_2}. \tag{A.4}$$

The similarities with, and differences from, the basic model of Chapter 11 are as we would expect.

During drug therapy ($\beta = 0$), wild-type virus will, as before, decline exponentially in the individual types of infected cells

$$y_i(t) = y_i^* e^{-a_i t}, \quad i = 1, 2, 3. \tag{A.5}$$

Free wild-type virus particles decay according to

$$v(t) = v^* e^{-ut} + y_1^* \frac{k}{u - a_1}(e^{-a_1 t} - e^{-ut}) + y_2^* \frac{c}{u - a_2}(e^{-a_2 t} - e^{-ut}). \tag{A.6}$$

If u is sufficiently larger than the other decay constants, then a good approximation is

$$v(t) = y_1^* \frac{k}{u}e^{-a_1 t} + y_2^* \frac{c}{u}e^{-a_2 t}. \tag{A.7}$$

We know that free virus declines with a half-life of about 2 days and hence the leading term of viral decay has to be the a_1-exponential decline. Therefore we conclude that most of the free plasma virus is produced by actively replicating cells. Latently infected cells can only contribute little to the plasma virus pool; in mathematical terms we must have $ky_1^* \gg cy_2^*$.

The full dynamics, including drug-resistant strains, is described by the following system:

$$\begin{aligned}
\dot{x} &= \lambda - dx - \beta xv - \beta_m xv_m, \\
\dot{y}_i &= q_i\beta(1 - \epsilon)xv - a_i y_i, \quad i = 1, 2, 3, \\
\dot{v} &= ky_1 + cy_2 - uv, \\
\dot{y}_{im} &= q_i\beta\epsilon xv + q_i\beta_m xv_m - a_i y_{im}, \quad i = 1, 2, 3, \\
\dot{v}_m &= k_m y_{1m} + c_m y_{2m} - uv_m.
\end{aligned} \tag{A.8}$$

For a small mutation rate, ϵ (and a basic reproductive ratio larger than one) the mutant virus and its infected cells are initially present at their low, mutation–selection levels, given by putting all time derivatives equal to zero in eqn (A.8). Neglecting relative order ϵ, we get:

$$y_{im}^* = \epsilon y_i^*/(1-f), \quad i = 1,2,3,$$

$$v_m^* = \epsilon v^*(\beta/\beta_m)f/(1-f). \tag{A.9}$$

Here we have, in analogy with the basic model, defined

$$f = \frac{R_m}{R_0}, \tag{A.10}$$

with R_0 defined by eqns (A.2) and (A.4), and R_m defined correspondingly as

$$R_m = \frac{\lambda\beta_m A_m}{du}. \tag{A.11}$$

Paralleling eqn (A.4), A_m is defined as

$$A_m = \frac{k_m q_1}{a_1} + \frac{c_m q_2}{a_2}. \tag{A.12}$$

The quantities x^*, v^*, y_i^* are defined as above, in eqn (A.3). These results are straightforward extensions of those for the earlier basic model.

We now turn to analyse the dynamics of the rise of resistant mutant virus populations, and their infected cells, following the sustained administration of drug after $t = 0$; that is, we look at eqn (A.8) with $\beta = 0$ for $t \geq 0$. For the following calculations we shall assume that most of the free virus is produced by the active rather than the latent cell pool. Thus, from now on, we neglect cy_2 compared to ky_1, and similarly for the mutant. This reasonable approximation will greatly reduce the complexity of our analysis, permitting insight into the numerical simulations.

With $\beta = 0$, and $c = c_m = 0$, the dynamics of the mutant virus and the cells it infects are given, for $t > 0$, by

$$\dot{y}_{im} = q_i(\beta_m k_m x y_{1m}/u) - a_i y_{im},$$

$$\dot{x} = \lambda - dx - (\beta_m k_m/u)y_{1m}x. \tag{A.13}$$

As before, we assume the viral turnover rate, u, is significantly faster than other rate constants, so that the mutant virus dynamics tracks that of y_{1m}:

$$v_m(t) \approx \left(\frac{k_m}{u}\right)y_{1m}(t). \tag{A.14}$$

It is apparent that the dynamics of $x(t)$ and $y_{1m}(t)$ are described by the pair of equations:

$$\dot{\hat{x}} = d[1 - \hat{x}(1 + \hat{y}_{1m})],$$

$$\dot{\hat{y}}_{1m} = a_1\hat{y}_{1m}[R_m\hat{x} - 1]. \tag{A.15}$$

The rescaling is exactly as in eqn (11.9) and R_m is defined by eqn (A.11) which, for $c_m = 0$, is

$$R_m = \frac{q_1 \lambda \beta_m k_m}{a_1 du}. \tag{A.16}$$

In eqn (A.15), the initial conditions are again given by eqn (11.11). The dynamics of the population of infected cells which are actively producing virus, $y_{1m}(t)$, and consequently the dynamics of the mutant virus population itself (via eqn (A.14)), again follow the several phases outlined in Chapter 11. Specifically, in phase I the population of infected cells which are actively producing viral mutants will increase as given by eqn (11.6):

$$\hat{y}_{1m}(t) = \hat{\epsilon} \, \exp[-(1 - f)a_1 t + \tfrac{1}{2}(R_m - f)a_1 d t^2 + \cdots]. \tag{A.17}$$

For the pool of cells latently infected with resistant mutant virus, $y_{2m}(t)$, we have the equation

$$\dot{y}_{2m}(t) = q_2(\beta_m k_m/u)x y_{1m} - a_2 y_{2m}. \tag{A.18}$$

This can be integrated to get

$$y_{2m}(t) = y_{2m}^* e^{-a_2 t} + (q_2 \beta_m k_m/u) \int_0^t x(s) y_{1m}(s) e^{-a_2(t-s)} \, ds. \tag{A.19}$$

On the RHS, the first term is initially of order ϵ, and thereafter decreases; the second term begins of order ϵ, but increases. Henceforth we neglect this first term on the RHS.

Similarly, from eqn (A.13) we obtain for the population of defectively infected cells the approximate expression

$$y_{3m}(t) = (q_3 \beta_m k_m/u) \int_0^t x(s) y_{1m}(s) e^{-a_3(t-s)} \, ds. \tag{A.20}$$

Alternatively, expressed in terms of the rescaled $\hat{x}$ and $\hat{y}_{1m}$ of eqn (A.15), we have

$$y_{3m}(t) = q_3 \lambda \int_0^t \hat{x}(s) \hat{y}_{1m}(s) e^{-a_3(t-s)} \, ds. \tag{A.21}$$

We conclude this section with remarks on the characteristic dynamics of each of the populations of cells, y_{1m}, y_{2m}, and y_{3m}. We refer both to absolute abundance, and— in analogy to the discussion of the basic model—to abundance relative to (declining) populations of corresponding cells infected with the wild-type virus. The discussion is illustrated with reference to the numerical simulations of Fig. A.1. *Our main purpose, however, is to gain qualitative insights which may help us extract information about vital*

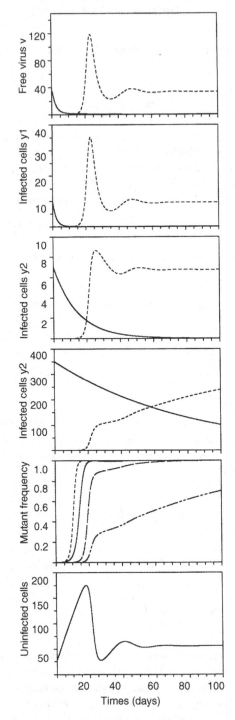

Fig. A.1 Computer simulation of the extended model (eqn A.8) describing the dynamics of drug-sensitive wild-type virus and drug-resistant mutant virus in the free

rates from empirical data on temporal abundance of uninfected and different types of infected cells.

(i) *Uninfected cells, $x(t)$.* With $\beta = 0$ for $t > 0$, we expect $x(t)$ to rise (at first roughly linearly, $x(t) \approx x^* + \lambda t(1 - 1/R_0) + O(t^2)$) during phase I. Eventually, $x(t)$ will damp to its asymptotic value at $x_m^* = 1/R_m$.

(ii) *Infected cells, actively producing resistant mutant virus, $y_{1m}(t)$.* Here we have eqn (A.17) as an approximate description of $y_{1m}(t)$ throughout 'phase I', where $\hat{y}_{1m} = (\beta_m k_m/du) y_{1m} < 1$. Thereafter, $y_{1m}(t)$ rises further on a fast, $1/a_1$, time scale, and subsequently falls and further executes damped oscillations, toward an asymptotic value of $y_{1m}(\infty) = (q_1\lambda/a_1)(1 - 1/R_m)$.

From a practical point of view, this suggests that we can estimate λ from the early (roughly linear) slope of the $x(t)$ versus t data, and thence estimate q_1/a_1 from the asymptotic value of $y_{1m}(t \to \infty)$ (assuming $1/R_m$ is smallish compared with unity).

For an estimate of $Y_1(t)$, the fraction of mutant-infected cells among the totality of actively-replicating-infected cells, we again have eqn (11.18)

$$Y_1(t) = \frac{\phi_1(t)}{\phi_1(t) + 1}. \tag{A.22}$$

Here $\phi_1(t) = y_{1m}(t)/y_1(t)$ is again given essentially by eqn (11.17) except a_1 replaces a, and R_m is given by eqn (A.16). The 50% point is thus again given by eqn (11.21)

$$\tfrac{1}{2}(R_m - f)a_1 dt_y^2 + fa_1 t_y - \ln[(1 - f)/\epsilon] = 0. \tag{A.23}$$

Thus, depending on what inferences can be drawn about parameter combinations by other means, observation of t_y from empirical data permits assessment of parameters like a_1, f, and ϵ.

For the specific parameter values chosen in the numerical simulations of Fig. A.1, we have $f = 0.5$, $R_m = 33.3$, $a_1 = 0.5$, $d = 0.01$ and $\epsilon = 0.0001$. Here the approximate eqn (11.22) gives $t_y \approx 10$ days, in good agreement with the computations illustrated in Fig. A.1.

---

virus population, v, and infected cell populations harbouring replication active virus, y_1, latent virus, y_2, and defective virus, y_3. The declining wild-type virus is shown by a continuous line, whereas the broken line indicates the rising mutant virus. The fifth panel from the top shows the relative frequency of mutant virus emerging in the actively infected cell population (broken line), followed by the free virus population (continuous line), followed by the latently infected cell population (broken line long dashes) and finally followed by the defectively infected cell population (broken line with two short dashes one long dash). The bottom panel displays the dynamics of the uninfected cell population, x. Parameter values are $\lambda = 10$, $d = 0.01$, $a_1 = 0.5$, $a_2 = 0.07$, $a_3 = 0.0125$, $u = 3$, $\beta = 0.01$, $\beta_m = 0.005$, $\epsilon = 0.0001$, $k = k_m = 10$ and $c = c_m = 1$ per day. The probabilities of producing actively infected, latently infected and defectively infected cells are $q_1 = 0.5$, $q_2 = 0.05$, and $q_3 = 0.45$, respectively. Our parameter choice implies that latently infected cells have an average life-span of about 14 days, whereas defectively infected cells have an average life-span of 80 days.

Notice also in this example that $k_m = k$, whence the mutant virus—expressed as a proportion of total virus—rises almost exactly as fast as do actively-replicating-infected cells. More generally, if $k_m < k$, we would see mutant virus, expressed proportionately (the V of eqn 11.19), rising later than actively-replicating-infected cells (the Y_1 defined above).

(iii) *Defectively infected cells,* $y_{3m}(t)$. Initially, the ratio $y_{3m}(0)/y_3(0)$ is $\epsilon/(1 - f)$. More generally, we can write the proportion of mutant-infected cells among all defectively-infected-cells as

$$Y_3(t) = \frac{y_{3m}}{y_{3m} + y_3} = \frac{\phi_3}{\phi_3 + 1}. \tag{A.24}$$

Here $\phi_3(t)$ is defined, from the good approximation of eqn (A.21), as

$$\phi_3(t) = \frac{a_3}{(1 - 1/R_0)} \int_0^t \hat{x}(s)\hat{y}_m(s)e^{a_3 s} \, ds. \tag{A.25}$$

Remember, $\hat{x}(0)\hat{y}_{1m}(0) = \epsilon f(1 - 1/R_0)/(1 - f)$.

For small t, ϕ_3 is of order ϵ. For very large t, $a_3 t \gg 1$, the non-exponential part of the integrand will asymptote to the value $\hat{x}(t)\hat{y}_{1m}(t) \to 1 - 1/R_m$; thence for $a_3 t \gg 1$ we get

$$\phi_3(t) \to \left(\frac{R_m - 1}{R_m - f}\right) e^{a_3 t}. \tag{A.26}$$

The passage from $\phi_3 \approx \epsilon$ to $\phi_3 \gg 1$ is, however, complex and dependent on the relative time-scales involved. We have already assumed that the turnover (death) rate of free virus is the fastest rate in the system ($u \gg a_i$ and d). We have also earlier assumed that virus-producing cells live significantly shorter than uninfected cells ($a_1 \gg d$). We now further explore the case where the death rate of defectively-infected cells, y_3, are comparable to uninfected ones, with latent cells, y_2, having intermediate death rates: $u \gg a_1 > a_2 \gg a_3 \approx d$.

These assumptions seem biologically reasonable. They appear to accord roughly with the available data on HIV infection, and they pertain to the numerical simulations of Fig. A.1. Under these assumptions, in the extreme limit where a_3 and d are so much slower than a_1 that we can regard $\hat{x}$ and $\hat{y}_{1m}$ as moving to their long-term equilibrium values effectively instantly (on an $1/a_3$-time-scale), we could just write $\phi_3 \approx a_3(R_m - 1)/(R_m - f) \int_0^t e^{a_3 s} \, ds$. The time for 50% of all defectively-infected cells to be those infected with mutant virus, t_{y3}, is then given via $Y_3 = 0.5$ and $\phi_3 = 1$ as

$$t_{y3} \approx \left(\frac{1}{a_3}\right) \ln\left[1 + \frac{R_m - f}{R_m - 1}\right]. \tag{A.27}$$

In the usual case where R_m is significantly in excess of unity, we have the estimate

$$t_{y3} \approx \frac{\ln 2}{a_3}. \tag{A.28}$$

In this event, empirical data which permit an estimate of t_{y3} lead directly to assessment of the turnover rate for defectively-infected cells, a_3. Notice, incidentally, that the approximate eqn (A.28) gives $t_{y3} \approx 55$ days for the parameter values in Fig. A.1, in excellent agreement with the numerical computations.

More generally, we will often have cases where, although a_3 is significantly larger than a_1, we cannot accurately regard $\hat{x}\hat{y}_{1m}$ to attain their asymptotic values effectively instantaneously in eqn (A.25). Turning back to the discussion following eqn (11.15), we can offer some rough insights. Phase I of the $\hat{x} - \hat{y}_{1m}$ dynamics lasts for a time roughly of order $1/R_m d$, and at the end of this phase $\hat{x}$ is of order $1/R_m$ and $\hat{y}_{1m}$ is of order 1. So we could say that after $t \approx 1/R_m d$, ϕ_3 is roughly of order (a_3/dR_m^2), or $1/R_m^2$ if a_3 and d are comparable. In phase II, $\hat{x}$ remains very roughly of order $1/R_m$, while $\hat{y}_{1m}$ gets significantly larger than unity; all this is on a timescale of order $1/a_1$. So the contribution to ϕ_3 from phase II is of general order of a_3/a_1, which is small. Overall, this points to the approximation of eqn (A.27) or (A.28) being reasonably accurate, so long as $a_3 \gg a_1$ and $R_m d$ (which implies the requirement that R_m be significantly larger than unity). Figure A.1 bears this out: the early stages of y_3 do show some of the features of y_1, but they settle relative quickly on the timescale of the y_3 dynamics.

(iv) *Latently infected cells*, y_{2m}. Here we have expressions for the absolute number of such cells infected with mutant virus, $y_{2m}(t)$, and for the proportion of these mutant virus infected cells among all such latent cells, $y_2(t) = y_{2m}/(y_2 + y_{2m})$. These expressions are exactly as in eqns (A.24)–(A.28) above, with all subscripts '3' replaced by '2'.

The analysis is, however, not usually as transparent as it can be for y_{1m} and y_{3m}. If a_2 is of the order of a_1, so that latent cells infected with mutant virus come to preponderate within the 'phase I' stage of the growth of y_{1m}, then analytic approximations to eqn (A.25) can be constructed; in essentials, the curve for $Y_2(t)$ behaves similarly to that for $Y_1(t)$. At the opposite extreme, if a_2 is of the order of a_3, then all the remarks just made about y_3 and y_{3m} pertain equally to y_2 and y_{2m}.

Most likely, however, a_2 will lie at values intermediate between a_1 and a_3, such that nothing simple can be said about the dynamics of $Y_2(t)$. This is the case for the parameter values illustrated in Fig. A.1.

(v) *Total abundance of infected cells*. Any interpretation of the dynamics of the total population of cells infected with mutant virus, $y_m(t) = y_{1m}(t) + y_{2m}(t) + y_{3m}(t)$, will depend significantly upon the factors which affect their initial relative abundances, and their asymptomatic relative abundances. For the total proportion of mutant-infected cells among all infected cells, $Y(t) = y_m(t)/[y(t) + y_m(t)]$, patterns will further depend on the initial relative abundances of the wild-type-infected cells. Insofar as a_3 is typically much smaller than a_1 and a_2, defective cells are likely to be the most abundant of wild-type infected cells at $t = 0$. These cells, moreover, decay slowly (as $e^{-a_3 t}$), and so the rise in $Y(t)$ is likely to be dominated by the asymptotic phase of $y_m(t)$; although $Y_1(t)$ saturates to unity relatively fast, by virtue of the fast decay in $y_1(t)$ (scaling as $e^{-a_1 t}$), this does not contribute much to the larger picture of all cells, dominated by the relatively large initial number and slow decay of y_3-cells. Such general observations again help us interpret the trends shown in the numerical simulations.

APPENDIX B

ANALYSIS OF MULTIPLE EPITOPE DYNAMICS

This appendix has four parts, each giving more technical detail in support of the analysis of the multiple epitope models in Chapter 14.

B.1 An invariant of motion

Consider the following Lotka–Volterra system of Chapter 14:

$$\dot{v}_{ij} = v_{ij}(r_{ij} - p_i x_i - q_j y_j),$$
$$\dot{x}_i = x_i(c_i v_{i*} - b_i) \quad \text{with } i = 1, \ldots, n_1, \tag{B.1}$$
$$\dot{y}_j = y_j(k_j v_{*j} - d_j) \quad \text{with } j = 1, \ldots, n_2.$$

This is eqn (14.11) in Section 14.3. The only difference is that we also allow for different natural decay rates, b_i and d_j, of the immune cells.

Let us consider a subsystem of x_i's, y_j's and v_{kl}'s, and assume that all other species are not present. We call this subsystem Γ a *solvable array* if the corresponding system of linear equations for the fixed point has some solution $\bar{x}_i$, $\bar{y}_j$ and $\bar{v}_{kl}$. More precisely, we require the following:

(i) for every x_i belonging to Γ, the set of all v_{ij} belonging to Γ is non-empty;
(ii) a corresponding condition for all y_j belonging to Γ;
(iii) for every v_{kl} belonging to Γ, there exists an x_k or an y_l belonging to Γ;
(iv) the corresponding set of linear equations

$$r_{kl} = x_k + y_l, \tag{B.2}$$

$$\frac{b_i}{c_i} = \sum_j v_{ij}, \tag{B.3}$$

$$\frac{d_j}{k_j} = \sum_i v_{ij} \tag{B.4}$$

has a solution $\bar{x}_i$, $\bar{y}_j$, $\bar{v}_{kl}$.

(Here we consider only those variables belonging to Γ. If, for instance, x_k does not belong to Γ, then the first equation reads $r_{kl} = y_l$. Similarly, the sum in the second equation extends over all those j for which v_{ij} belongs to Γ. We do not require that all these quantities are positive or uniquely determined.)

Whenever we have such a solvable array, the function

$$V = \sum_{kl}(\bar{v}_{kl} \log v_{kl} - v_{kl}) + \sum_i \frac{1}{c_i}(\bar{x}_i \log x_i - x_i) + \sum_j \frac{1}{k_j}(\bar{y}_j \log y_j - y_j)$$

(B.5)

is an invariant of motion. To prove this, we note that

$$\dot{V} = \sum_{kl}(\bar{v}_{kl} - v_{kl})(r_{kl} - x_k - y_l) + \sum_i(\bar{x}_i - x_i)\left(\sum_j v_{ij} - \frac{b_i}{c_i}\right)$$
$$+ \sum_j(\bar{y}_j - y_j)\left(\sum_i v_{ij} - \frac{d_j}{c_j}\right).$$

(B.6)

Upon replacing r_{kl} by $\bar{x}_k + \bar{y}_l$, b_i/c_i by $\sum \bar{v}_{ij}$, and d_j/K_j by $\sum \bar{v}_{ij}$, we obtain

$$\dot{V} = \sum_{kl}(\bar{v}_{kl} - v_{kl})(\bar{x}_k - x_k + \bar{y}_l - y_l) + \sum_i(\bar{x}_i - x_i)\sum_j(v_{ij} - \bar{v}_{ij})$$
$$+ \sum_j(\bar{y}_j - y_j)\sum_i(v_{ij} - \bar{v}_{ij}).$$

(B.7)

That is,

$$\dot{V} = \sum_i(\bar{x}_i - x_i)\sum_j(\bar{v}_{ij} - v_{ij} + v_{ij} - \bar{v}_{ij})$$
$$+ \sum_l(\bar{y}_l - y_l)\sum_k(\bar{v}_{kl} - v_{kl} + v_{kl} - \bar{v}_{kl}),$$

(B.8)

which reduces to $\dot{V} = 0$.

If the fixed point given by the $(\bar{x}_i, \bar{y}_j$ and $\bar{v}_{kl})$ has all components positive, then the function V attains its unique maximum at this point. Hence this equilibrium is neutrally stable and all eigenvalues are purely imaginary. All populations originally present (i.e. belonging to the array Γ) persist forever. However, the system is not permanent: a sequence of random perturbations can send the state from one-level set to another, and thus eventually to the boundary of the positive state space. We have seen that we can have at most $n_1 + n_2 - 1$ viral species present in a solvable array. If, on the other hand, some components of the fixed point are negative, the corresponding populations have to vanish (possibly after an initial phase of growth).

B.2   Local dynamics of a multiple epitope equation

In this appendix, we give a linearized analysis of the dynamics of the system in which activated CTLs arise from inactivated precursors, eqn (14.20), in the biologically interesting limit when all r_{ij} have the same value, $r_{ij} = r$.

In the usual way, we begin by writing

$$x_i(t) = \xi + v_i(t), \tag{B.9}$$
$$y_j(t) = (1 - \xi) + \phi_j(t), \tag{B.10}$$
$$v_{ij}(t) = v_{ij}^* + \chi_{ij}(t). \tag{B.11}$$

Here v_i, ϕ_j, and χ_{ij} represent small perturbations about the interior equilibrium defined by eqns (14.23) and (14.24). As discussed in the main text, the individual equilibrium values of v_{ij} (here denoted by v_{ij}^*) can take arbitrary values within the envelope set by the equilibrium values of v_{i*} and v_{*j}, as given by eqn (14.24). We now substitute eqns (B.9)–(B.11) into eqn (14.20), Taylor expand to first order (discarding all terms of second or higher order in v_i, ϕ_i, and χ_{ij}), and factor out the time-dependence in the ensuing set of linearized differential equations as $\exp(\Lambda t)$:

$$\Lambda \chi_{ij} = -v_{ij}^*(v_i + \phi_j), \tag{B.12}$$

$$\Lambda v_i = c_i(\eta p_i + \xi) \sum_{k=1}^{n_2} \chi_{ik} + (c_i v_{i*} - b)v_i, \tag{B.13}$$

$$\Lambda \phi_j = k_j(\eta q_j + 1 - \xi) \sum_{k=1}^{n_1} \chi_{kj} + (k_j v_{*j} - b)\phi_j. \tag{B.14}$$

Using eqn (14.24) to substitute for the equilibrium values of v_{i*} and v_{*j}, and using eqn (B.12) to substitute for χ_{ij} in eqns (B.13) and (B.14), we arrive at a set of $n_1 + n_2$ linear equations for the (small) perturbations to x_i and y_j, namely v_i and ϕ_j, respectively:

$$\left(\Lambda + \frac{\eta p_i b}{\eta p_i + \xi}\right)v_i + \frac{c_i}{\Lambda}(\eta p_i + \xi) \sum_{l=1}^{n_2} v_{il}^*(v_i + \phi_l) = 0, \tag{B.15}$$

$$\left(\Lambda + \frac{\eta q_j b}{\eta q_j + 1 - \xi}\right)\phi_i + \frac{k_j}{\Lambda}(\eta q_j + 1 - \xi) \sum_{h=1}^{n_1} v_{hj}^*(v_h + \phi_j) = 0. \tag{B.16}$$

Rearrangement, and further use of eqn (14.24), gives:

$$\left[\Lambda^2 + \Lambda \frac{\eta p_i b}{\eta p_i + \xi} + b\xi\right]v_i + c_i(\eta p_i + \xi) \sum_{l=1}^{n_2} v_{il}^*\phi_l = 0, \tag{B.17}$$

$$\left[\Lambda^2 + \Lambda \frac{\eta q_j b}{\eta q_j + 1 - \xi} + b(1 - \xi)\right]\phi_j + k_j(\eta q_j + 1 - \xi) \sum_{h=1}^{n_1} v_{hj}^* v_h = 0. \tag{B.18}$$

Equations (B.17) and (B.18) represent a homogeneous, linear set of equations for the $n_1 + n_2$ variables $\{v_i\}$ and $\{\phi_j\}$. The corresponding $(n_1 + n_2) \times (n_1 + n_2)$ matrix of coefficients must therefore have a vanishing determinant (note that this matrix partitions into two purely diagonal submatrices, $n_1 \times n_1$ and $n_2 \times n_2$, and two other submatrices,

$n_1 \times n_2$ and $n_2 \times n_1$, whose elements depend upon the arbitrary (subject to constraints) values of v_{ij}^*). The requirement that this overall determinant be zero leads to values for the quantities Λ which characterize the time-dependence, and hence to elucidation of the system's local stability properties. We have not succeeded in showing that $\mathrm{Re}(\Lambda) \leq 0$ for the general case of eqns (B.17) and (B.18), but we can make some progress in the special case when p_i and q_j are constants ($p_i = p$, $q_j = q$).

In this case, eqns (B.17) and (B.18) can be reduced to the form

$$\sum_{h=1}^{n_1} [A_{ih} - F(\Lambda)\delta_{ih}]v_h = 0. \tag{B.19}$$

Here δ_{ih} is the Kronecker delta, and the $n_1 \times n_1$ matrix A has elements

$$A_{ih} = c_i \sum_{l=1}^{n_2} k_l v_{il}^* v_{hl}^*. \tag{B.20}$$

The function $F(\Lambda)$ is defined as

$$F(\Lambda) = \left[\Lambda^2 + \Lambda \frac{\eta p b}{\eta p + \xi} + b\xi \right] \left[\Lambda^2 + \Lambda \frac{\eta q b}{\eta q + 1 - \xi} + b(1 - \xi) \right]$$
$$\times [(\eta p + \xi)(\eta q + 1 - \xi)]^{-1}. \tag{B.21}$$

Denote the eigenvalues of the matrix A as λ_i ($i = 1, 2, \ldots, n_1$). The stability-determining quantities Λ are then found by solving the quartic equations

$$F(\Lambda) = \lambda_i. \tag{B.22}$$

Notice that A is symmetric, up to the row-constants c_i, whence it follows that all the eigenvalues λ_i are real (see, e.g., May 1973).

In general, the eigenvalues of the matrix A defined by eqn (B.20) depend on the values of $\{v_{ij}^*\}$, and cannot be obtained analytically. We can, however, get exact solutions in two limiting cases, which are likely to 'bracket' more general cases.

One limiting case arises when all v_{ij}^* are equal (this is, indeed, the asymptotic result which seems to occur in most of our numerical simulations with $p_i = p$, $q_i = q$ and $r_{ij} = 1$). In this case, we can use eqn (14.24) to write $c_i v_{il}^* = b\xi/[n_2(\eta p + \xi)]$ and $k_l v_{hl}^* = b(1-\xi)/[n_1(\eta q+1-\xi)]$. Then all elements of the matrix A have the same value, $A_{ik} = a \equiv b^2\xi(1-\xi)/[n_1(\eta p+\xi)(\eta q+1-\xi)]$. Such an ($n_1 \times n_2$) matrix has $n_1 - 1$ eigenvalues $\lambda_i = 0$ ($i = 2, 3, \ldots, n_1$), and one eigenvalue $\lambda_1 = n_1 a$. Returning to eqn (B.22), we see that there are $n_1 - 1$ internal modes whose dynamics are characterized by Λ-values which obey $F(\Lambda) = 0$, with $F(\Lambda)$ the product of two quadratics, defined by eqn (B.21). All four Λ-values for each of these $n_1 - 1$ (identical) internal nodes then clearly lie in the left-half plane. Moreover, for small values of η, these two quadratics each correspond to weakly damped oscillations, two with frequency $\sqrt{b\xi}$ and characteristic damping time $2\xi/(\eta pb)$, and two with frequency $\sqrt{b(1 - \xi)}$ and characteristic damping time $2(1-\xi)/(\eta q b)$. The remaining Λ-values correspond to the dynamics of the system as a whole, and are given by $F(\Lambda) = n_1 a$, which reduces to

$$\left[\Lambda^2 + \Lambda\frac{\eta p b}{\eta p + \xi} + b\xi\right]\left[\Lambda^2 + \Lambda\frac{\eta q b}{\eta q + 1 - \xi} + b(1 - \xi)\right] = b^2\xi(1 - \xi). \quad \text{(B.23)}$$

The constant terms cancel, giving a cubic in Λ, of the form $\Lambda^3 + \alpha\Lambda^2 + \beta\Lambda + \gamma = 0$; it is easy to see that $\alpha > 0$, $\beta > 0$, $\gamma > 0$, and $\alpha\beta > \gamma$, so that all Λ-values lie in the left-half plane. In the limit of very small η, we get weakly damped oscillations with frequency $\sqrt{b}$ and characteristic damping time $2/[\eta b(p + q)]$; there is also a monotonically damped mode with damping time, τ, of around $\tau^{-1} = \eta b\{[p(1 - \xi)/\xi] + [q\xi/(1 - \xi)]\}$.

An opposite limiting case arises when $n_1 = n_2$ and each row and column of the v_{ij}^*-matrix has only one non-zero entry. In this case we can bring the v_{ij}^*-matrix into diagonal form, and thence write the eigenvalues of the matrix A as

$$\lambda_i = \frac{b^2\xi(1 - \xi)}{(\eta p + \xi)(\eta q + 1 - \xi)}, \quad \text{for all } i. \quad \text{(B.24)}$$

Substituting this into eqn (B.22) leads again to eqn (B.23) for all the Λ-values in this case. As above, all these stability-determining Λ-values lie in the left-half plane. Again we have weakly damped oscillations with frequencies $\sqrt{b}$ and characteristic damping times of order $1/\eta$ if η is very small.

It seems reasonable to assume that other assignments of $\{v_{ij}^*\}$, within the overall constraints set by the v_{i*} and v_{*j} of eqn (14.24), will lead to dynamics whose qualitative behaviour is bracketed by these two limiting cases. We thus expect the interior equilibrium of Section 14.4.1 generally to be locally stable.

B.3   The 2×2 system

B.3.1 $\eta = 0$

Let us now consider the system

$$\begin{aligned}
\dot{v}_{ij} &= v_{ij}(r_{ij} - x_i - y_j), \\
\dot{x}_i &= x_i(c_i v_{i*} - b) \quad \text{with } i = 1, 2, \\
\dot{y}_j &= y_j(k_j v_{*j} - b) \quad \text{with } j = 1, 2.
\end{aligned} \quad \text{(B.25)}$$

Note that the ratio $\rho = v_{11}v_{22}/(v_{12}v_{21})$ is a Lyapunov function:

$$\dot{\rho} = \rho(r_{11} + r_{22} - r_{12} - r_{21}). \quad \text{(B.26)}$$

If $r_{11} + r_{22} > r_{12} + r_{21}$, then $\rho \to \infty$ which implies that v_{12} or v_{21} (or both) have to converge to zero. Of course, this excludes the possibility of an interior equilibrium. Note that $\rho \to \infty$ does not exclude the possibility that also v_{11} or v_{22} may converge to zero.

We shall now give a full classification of system (B.25) for a generic choice of parameters. We shall show that the system always admits a unique saturated equilibrium

P, which lies on some boundary face (either one or two of the four viral species, and the same number of the CTL species, have to vanish). Within the corresponding boundary face, however, we know from Section B.1 that P is neutrally stable: all eigenvalues are on the imaginary axis, and the orbits do neither converge toward P nor diverge away from P. For every initial condition, the orbit converges towards the face defined by P. Those components which do not vanish will exhibit undamped oscillations. Their time-averages will be given by P.

Let us start the classification by assuming that

$$\frac{1}{c_1} + \frac{1}{c_2} < \frac{1}{k_1} + \frac{1}{k_2}. \tag{B.27}$$

This is no restriction of generality (if the converse inequality is valid, we just have to interchange x and y), and it implies, as we have seen, that at least one of the y-responses converges to 0. Next, we assume that

$$r_{12} + r_{21} < r_{11} + r_{22}. \tag{B.28}$$

Again, this can be achieved without restricting generality: if the converse inequality holds, we just have to exchange v_{11} with v_{12}, and v_{21} with v_{22}. These conditions imply that at least one of the viral species v_{12} and v_{21} vanishes. The remaining part of the parameter space will be divided into three mutually exclusive cases:

(A) $r_{11} < r_{12}$ (which implies $r_{21} < r_{22}$);
(B) $r_{12} < r_{11}$ and $r_{22} < r_{21}$;
(C) $r_{12} < r_{11}$ and $r_{21} < r_{22}$.

Each of these can be subdivided into three cases in turn:

(A1) $\frac{1}{k_2} < \frac{1}{c_2}$. In this case $y_1 = v_{12} = 0$.
(A2) $\frac{1}{c_2} < \frac{1}{k_2} < \frac{1}{c_1} + \frac{1}{c_2}$. In this case $y_1 = v_{21} = 0$.
(A3) $\frac{1}{c_1} + \frac{1}{c_2} < \frac{1}{k_2}$. In this case $y_1 = y_2 = v_{11} = v_{21} = 0$.

(B1) $\frac{1}{k_1} < \frac{1}{c_1}$. In this case $y_2 = v_{21} = 0$.
(B2) $\frac{1}{c_1} < \frac{1}{k_1} < \frac{1}{c_1} + \frac{1}{c_2}$. In this case $y_2 = v_{12} = 0$.
(B3) $\frac{1}{c_1} + \frac{1}{c_2} < \frac{1}{k_1}$. In this case $y_1 = y_2 = v_{12} = v_{22} = 0$.

(C1) $\frac{1}{k_2} < \frac{1}{c_2}$ (which implies $\frac{1}{c_1} < \frac{1}{k_1}$). In this case $y_1 = v_{12} = 0$.
(C2) $\frac{1}{c_2} < \frac{1}{k_2}$ and $\frac{1}{k_1} < \frac{1}{c_1}$. In this case $y_2 = v_{21} = 0$.
(C3) $\frac{1}{c_2} < \frac{1}{k_2}$ and $\frac{1}{c_1} < \frac{1}{k_1}$. In this case $y_1 = y_2 = v_{12} = v_{21} = 0$.

All other components of P are strictly positive and can easily be computed. It is a straightforward, but rather tedious, task to check that in each case no equilibrium other than P is saturated.

As an example, let us consider the fixed point in the interior of the face ($y_2 = 0$, $v_{22} = 0$) which is given by

$$v_{21}^* = \frac{b}{c_2}, \quad v_{11}^* = \frac{b}{k_1} - \frac{b}{c_2}, \quad v_{12}^* = \frac{b}{c_1} + \frac{b}{c_2} - \frac{b}{k_1},$$

$x_1^* = r_{12}$, $y_1^* = r_{11} - r_{12}$ and $x_2^* = r_{21} + r_{12} - r_{11}$. Since these quantities have to be positive, we must have

$$r_{12} < r_{11} < r_{12} + r_{21} \tag{B.29}$$

and

$$\frac{1}{c_2} < \frac{1}{k_1} < \frac{1}{c_1} + \frac{1}{c_2}. \tag{B.30}$$

If the parameters are chosen properly, none of the missing species y_2 and v_{22} can invade. Indeed, the fixed point is saturated in the sense that the two transversal eigenvalues $\dot{y}_2/y_2$ and $\dot{v}_{22}/v_{22}$ are negative. These eigenvalues are given by $r_{22} - x_2^* = r_{22} + r_{11} - r_{12} - r_{21}$ and by $k_2 v_{12}^* - d_2$, which is a positive multiple of

$$\frac{1}{c_2} + \frac{1}{c_2} - \frac{1}{k_1} - \frac{1}{k_2}.$$

Thus, we have to choose

$$r_{11} + r_{22} < r_{12} + r_{21} \tag{B.31}$$

and

$$\frac{1}{c_1} + \frac{1}{c_2} < \frac{1}{k_1} + \frac{1}{k_2}. \tag{B.32}$$

If these conditions are satisfied, and we start with the full system (i.e. all eight populations positive), then y_2 and v_{22} vanish and the remaining species will persist.

Furthermore, note that the function

$$V = (v_{11}^* \log v_{11} - v_{11}) + (v_{12}^* \log v_{12} - v_{12}) + (v_{21}^* \log v_{21} - v_{21})$$
$$+ \frac{1}{c_1}(x_1^* \log x_1 - x_1) + \frac{1}{c_2}(x_2^* \log x_2 - x_2) + \frac{1}{k_1}(y_1^* \log y_1 - y_1) \tag{B.33}$$

is a constant of motion. All orbits lie on the constant level sets in V. This implies that within its face the fixed point is neutrally stable. All saturated fixed points of the 2×2 system defined by eqn (B.25) are listed in Table B.1.

B.3.2 $\eta > 0$

If we now consider the case of small $\eta > 0$, we see that its saturated equilibria points must, by continuity, converge (for $\eta \to 0$) to saturated equilibria of the $\eta = 0$ case. Hence P is the only possible candidate; it follows that at least for small for $\eta > 0$, the system has a unique saturated fixed point. This point differs from P by having small, but positive, values for those CTL's which, in the $\eta = 0$-equilibrium, were not present but have viral species which stimulate their replication. The pattern of virus distribution,

Table B.1 *Saturated fixed points of the 2 × 2 system with $\eta = 0$, as specified by eqn (B.25)*

v_{ij} matrix	(x_1, x_2)	(y_1, y_2)	Case	Conditions of existence and stability
$\begin{pmatrix} + & 0 \\ + & + \end{pmatrix}$	$(+,+)$	$(0,+)$	A1	$r_{11} < r_{12}$ $1/k_2 < 1/c_2$ $r_{21} < r_{22}$
			C1	$r_{11} > r_{12}$ $1/k_2 < 1/c_2$ $r_{21} < r_{22}$
$\begin{pmatrix} + & 0 \\ + & + \end{pmatrix}$	$(+,+)$	$(+,0)$	B2	$r_{11} > r_{12}$ $1/c_1 < 1/k_1 < 1/c_1 + 1/c_2$ $r_{21} > r_{22}$
$\begin{pmatrix} + & + \\ 0 & + \end{pmatrix}$	$(+,+)$	$(0,+)$	A2	$r_{11} < r_{12}$ $1/c_2 < 1/k_2 < 1/c_1 + 1/c_2$ $r_{21} < r_{22}$
$\begin{pmatrix} + & + \\ 0 & + \end{pmatrix}$	$(+,+)$	$(+,0)$	B1	$r_{11} > r_{12}$ $1/k_1 < 1/c_1$ $r_{21} > r_{22}$
			C2	$r_{11} > r_{12}$ $1/k_1 < 1/c_1$ $r_{21} < r_{22}$
$\begin{pmatrix} 0 & + \\ 0 & + \end{pmatrix}$	$(+,+)$	$(0,0)$	A3	$r_{11} < r_{12}$ $1/c_1 + 1/c_2 < 1/k_2$ $r_{21} < r_{22}$
$\begin{pmatrix} + & 0 \\ + & 0 \end{pmatrix}$	$(+,+)$	$(0,0)$	B3	$r_{11} > r_{12}$ $1/c_1 + 1/c_2 < 1/k_1$ $r_{21} > r_{22}$
$\begin{pmatrix} + & 0 \\ 0 & + \end{pmatrix}$	$(+,+)$	$(0,0)$	C3	$r_{11} > r_{12}$ $1/c_1 < 1/k_1$ and $1/c_2 < 1/k_2$ $r_{21} < r_{22}$

Without loss of generality we have assumed $1/c_1 + 1/c_2 < 1/k_1 + 1/k_2$ (which implies that at least one y_i has to converge to zero) and $r_{12} + r_{21} < r_{11} + r_{22}$ (which implies that v_{12} or v_{21}—or both—have to converge to zero). There are seven stable fixed points, characterised by nine parameter regions. Each parameter region admits exactly one stable fixed point. The conditions A3, B3, and C3 specify the interesting situation of complete immunodominance (i.e. $y_1 = y_2 = 0$). Note that this lack of y-responses can either occur with homogeneity (A3, B3) or heterogeneity (C3) in the y-epitope.

on the other hand, remains unchanged, since the v-equations do not depend on η. Hence with small $\eta > 0$, the number of CTL species will be higher, by 1 or 2, than the number of viral species.

A complete classification of the 2 × 2 case for general $\eta > 0$ is not possible, but below we give a complete analysis for the interesting limit of large η. There we find 10 mutually exclusive parameter regions which cover the whole parameter space like a jigsaw puzzle. Each parameter region specifies exactly one stable fixed point. Therefore by continuity we conjecture that also the general η system admits always a single stable fixed point.

B.3.3 The limit of large η

In this 2×2 case, we give a complete listing of all 10 saturated fixed points, and we show that one, and only one, of these 10 states exists for any specific choice of the parameters. Table B.2 shows how to determine which state, dependent upon 13 inequalities among the parameters, as defined below. This illustrative example requires only that η be large, in

Table B.2 *Saturated fixed points of the system of eqn (14.26), for the case $n_1 = n_2 = 2$, in the limit of large η*

State specified by v_{ij} matrix (+ represents a positive value of v_{ij})	Number labelling the state	Inequalities to be satisfied for this state to be saturated fixed point
$\begin{pmatrix} 0 & + \\ + & + \end{pmatrix}$	1	$\bar{A}, B_{22}, C_{22}, D_{22}$
$\begin{pmatrix} + & 0 \\ + & + \end{pmatrix}$	2	$A, B_{21}, C_{21}, D_{21}$
$\begin{pmatrix} + & + \\ 0 & + \end{pmatrix}$	3	$A, B_{12}, C_{12}, D_{12}$
$\begin{pmatrix} + & + \\ + & 0 \end{pmatrix}$	4	$\bar{A}, B_{11}, C_{11}, D_{11}$
$\begin{pmatrix} 0 & + \\ + & 0 \end{pmatrix}$	5	$\bar{A}, \bar{B}_{11}, \bar{B}_{22}$
$\begin{pmatrix} + & 0 \\ 0 & + \end{pmatrix}$	6	$A, \bar{B}_{12}, \bar{B}_{21}$
$\begin{pmatrix} + & 0 \\ + & 0 \end{pmatrix}$	7	$\bar{C}_{11}, \bar{C}_{21}$
$\begin{pmatrix} 0 & + \\ 0 & + \end{pmatrix}$	8	$\bar{C}_{22}, \bar{C}_{12}$
$\begin{pmatrix} + & + \\ 0 & 0 \end{pmatrix}$	9	$\bar{D}_{11}, \bar{D}_{12}$
$\begin{pmatrix} 0 & 0 \\ + & + \end{pmatrix}$	10	$\bar{D}_{22}, \bar{D}_{21}$

The symbols $A, B_{ij}, C_{ij}, D_{ij}$ stand for the inequalities (B.46)–(B.49), and the 'bars' denote the opposite inequality.

the sense defined below; a completely general analysis of the 2×2 case is not feasible, although intuition backed by numerical studies suggests that for any specified set of parameter values there will in general be a unique saturated fixed point.

For the 2×2 case of the system given by eqn (14.26) in Section 14.4.3, equilibrium values of the four variables v_{ij} are found by putting $\dot{v}_{ij} = 0$, which—as discussed in the main text—gives *either* $r_{ij} = x_i + y_j$ *or* $v_{ij} = 0$ (along with the condition $r_{ij} < x_i + y_j$). In the remaining eqn (14.26), setting $\dot{x}_i = 0$ and $\dot{y}_j = 0$ leads to the further conditions $v_{i*} = bx_i/(\eta c_i p_i + c_i x_i)$ and $v_{*j} = by_j/(\eta k_j q_j + k_j y_j)$, which in combination with the earlier equations involving r_{ij} lead to a complete specification.

There are thus two possible solutions for each v_{ij}, leading to $2^4 = 16$ possible solutions in total. But, as discussed in the main text, we cannot have saturated fixed points for which an entire row or column of the v_{ij} matrix vanish: this rules out the single solution where all $v_{ij} = 0$, and the four where only one $v_{ij} \neq 0$. Also, as noted in Section 14.2, the solution with all four of the $v_{ij} \neq 0$ is not generically possible. This leads to 10 cases to be examined, four with one $v_{ij} = 0$ and the other three non-zero, and six with two of the $v_{ij} = 0$ and the other two non-zero.

As an example, we sketch the derivation of the condition for the existence of a saturated fixed point with $v_{11} = 0$ and $v_{ij} \neq 0$ otherwise. These conditions correspond to particular inequalities that the parameters $\{r_{ij}\}$, $\{p_i c_i\}$, and $\{q_j k_j\}$ must satisfy. Without discussing the other nine cases in detail, we then set out conditions for the existence of each of the 10 possible saturated fixed points. This is done in Table B.2.

Finally, we emphasize that for a specified set of parameter values, one and only one of the 10 states of Table B.2 will arise. That is, the patchwork of inequalities summarized in Table B.2 fits together like a jigsaw puzzle. This result is not immediately obvious, and we conclude this appendix by sketching the proof.

The illustrative case of $v_{11} = 0$. For $v_{11} = 0$ and saturated, we require $\dot{v}_{11} < 0$, which implies $r_{11} < x_1 + y_1$. From $v_{ij} \neq 0$ for $i, j \neq 1, 1$, we have the three equations $r_{ij} = x_i + y_j$ when $i, j \neq 1, 1$. There are also four relations among v_{ij} and x_i, y_j, as follows:

$$v_{12} = \frac{bx_1}{\eta c_1 p_1 + c_1 x_1}, \tag{B.34}$$

$$v_{21} + v_{22} = \frac{bx_2}{\eta c_2 p_2 + c_2 x_2}, \tag{B.35}$$

$$v_{21} = \frac{by_1}{\eta k_1 q_1 + k_1 y_1}, \tag{B.36}$$

$$v_{12} + v_{22} = \frac{by_2}{\eta k_2 q_2 + k_2 y_2}. \tag{B.37}$$

As emphasized above, the analysis in this appendix depends (only) on the limiting assumption that η is large, in the sense that the second term in the brackets in each of the eqns (B.34)–(B.37) can be ignored; effectively, this means $\eta \gg$ (terms of order of $r_{ij}/p_i, q_j$). In this limit, we have a set of linear relations between v_{12}, v_{21}, v_{22} and x_i,

y_j. Combining these with the three equations $r_{ij} = x_i + y_j$ $(i, j \neq 1, 1)$, we have:

$$br_{12} = \eta(c_1 p_1 + k_2 q_2)v_{12} + \eta k_2 q_2 v_{22}, \tag{B.38}$$

$$br_{21} = \eta(c_2 p_2 + k_1 q_1)v_{21} + \eta c_2 p_2 v_{22}, \tag{B.39}$$

$$br_{22} = \eta k_2 q_2 v_{12} + \eta c_2 p_2 v_{21} + \eta(c_2 p_2 + k_2 q_2)v_{22}. \tag{B.40}$$

Solving this set of linear equations gives explicit expressions for v_{12}, v_{21}, v_{22}, and thence for x_1, x_2 and y_1, y_2.

By tedious but routine algebraic manipulations, it can be seen that the conditions $v_{12} > 0$, $v_{21} > 0$, $v_{22} > 0$ lead, respectively, to the requirements:

$$r_{22} < r_{12}\left[1 + \frac{v_2 \rho_1}{\rho_2(v_2 + \rho_1)}\right] + r_{21}\left(\frac{v_2}{v_2 + \rho_1}\right), \tag{B.41}$$

$$r_{22} < r_{12}\left(\frac{\rho_2}{v_1 + \rho_2}\right) + r_{21}\left[1 + \frac{v_1 \rho_2}{v_2(v_1 + \rho_2)}\right], \tag{B.42}$$

$$r_{22} > r_{12}\left(\frac{\rho_2}{v_1 + \rho_2}\right) + r_{21}\left(\frac{v_2}{v_2 + \rho_1}\right). \tag{B.43}$$

Here we have, for notational convenience, defined

$$v_i = c_i p_i \quad \text{and} \quad \rho_j = k_j q_j. \tag{B.44}$$

Clearly x_i and y_j are all positive if v_{12}, v_{21}, v_{22} are. The remaining requirement is that $x_1 + y_1 > r_{11}$ (so that $\dot{v}_{11} < 0$), which immediately implies the inequality

$$r_{11} + r_{22} < r_{12} + r_{21}. \tag{B.45}$$

Listing the 10 possible states These calculations can obviously be repeated, *mutatis mutandis*, for the other three possible states with a single $v_{ij} = 0$, and for the six states with two of the $v_{ij} = 0$. These 10 states, and the inequalities which must be satisfied for each of them to be a saturated fixed point, are catalogued in Table B.2. In this table, the symbols A, B_{ij}, C_{ij}, D_{ij} $(i, j = 1, 2)$ refer to the following inequalities:

$$A: \quad r_{11} + r_{22} > r_{12} + r_{21}, \tag{B.46}$$

$$B_{ij}: \quad r_{ij} > r_{Ij}[\rho_j/(v_I + \rho_j)] + r_{iJ}[v_i/(v_i + \rho_J)], \tag{B.47}$$

$$C_{ij}: \quad r_{ij} < r_{Ij}[\rho_j/(v_I + \rho_j)] + r_{iJ}[1 + v_I \rho_j/\{v_i(v_I + \rho_j)\}], \tag{B.48}$$

$$D_{ij}: \quad r_{ij} < r_{Ij}[1 + v_i \rho_J/\{\rho_j(v_i + \rho_J)\}] + r_{iJ}[v_i/(v_i + \rho_J)]. \tag{B.49}$$

Here the capital letter subscripts denote $I = 2$ if $i = 1$, and conversely. If the inequalities do not hold (i.e., if we have the opposite), we write $\bar{A}$, $\bar{B}_{ij}$, etc.

Uniqueness It can be seen that the set of inequalities listed in the rightmost column of Table B.2 imply that one, and only one, state will ensue for any arbitrary choice of the underlying parameters $\{r_{ij}\}$, $\{v_i\}$, $\{\rho_j\}$ which determine the inequalities A, B_{ij}, C_{ij}, D_{ij}.

The proof is straightforward, but a bit intricate, and depends on relations among the inequalities themselves. Thus it can be seen that $\bar{A} \wedge B_{11} \Rightarrow \bar{B}_{22} \Rightarrow C_{22}, D_{22}$; $A \wedge B_{12} \Rightarrow \bar{B}_{21} \Rightarrow C_{21}, D_{21}$; and so on ($\wedge$ stands for and). Likewise, although with a bit more difficulty, it can be shown that $\bar{C}_{22} \wedge \bar{D}_{22} \Rightarrow r_{22} > r_{12} + r_{21} \Rightarrow A$; $\bar{C}_{12} \wedge \bar{D}_{12} \Rightarrow \bar{A}$; etc. Finally, it can also be shown that $A \wedge \bar{C}_{ii} \Rightarrow \bar{C}_{Ii}$; $A \wedge \bar{C}_{Ii} \Rightarrow \bar{C}_{ii}$; $\bar{A} \wedge \bar{D}_{ii} \Rightarrow \bar{D}_{iI}$; and $A \wedge \bar{D}_{iI} \Rightarrow \bar{D}_{ii}$. Threading our way through the resulting maze, we find there is one, and only one, state for each specified set of parameters (and consequent inequalities).

We conclude with one example, to make these ideas more concrete. Suppose the values of $r_{ij}, c_i, k_j, p_i, q_j$ are such that $r_{12} + r_{21} > r_{11} + r_{22}$ ($\bar{A}$), and also that B_{22} is satisfied. This implies $\bar{B}_{11}$, and thence C_{11} and D_{11}. Working down the right-hand column of Table B.2, we see that the only possible states are then 1, 8, 10. If C_{22} and D_{22} are both satisfied, then we have the unique answer of state 1. If $\bar{C}_{22}$, then the interrelations listed above imply $\bar{C}_{12}$ (and also D_{22}), whence state 8 is the only answer. Conversely, if $\bar{D}_{22}$ then $\bar{D}_{21}$ (and also C_{22}), so that state 10 is the unique answer.

Systematic elaboration of these lines of argument leads to the conclusion that any choice of parameters leads (in the limit of large η) to one, and only one, of the 10 states catalogued in Table B.2.

B.4 Intracellular competition between epitopes

This appendix shows that the set of equations (14.42) of Section 14.7, which describe multiple epitope dynamics with intracellular competition between epitopes, cannot have a locally stable interior equilibrium.

As in Appendix B.2 (eqns (B.9)–(B.11)), we expand the variables about the interior fixed point (given by eqn (14.43)): $x_i(t) = \xi + v_i(t)$, $y_j(t) = \xi' + \phi_j(t)$, $v_{ij}(t) = v_{ij}^* + \chi_{ij}(t)$ (here $\xi' \equiv r - \xi$). We now Taylor-expand eqn (14.42) to first order, and factor out the time-dependence in the ensuing set of linear equations as $\exp(\Lambda t)$, to get:

$$\Lambda \chi_{ij} = -v_{ij}^*(v_i + \phi_j), \tag{B.50}$$

$$\Lambda v_i = \sum_{l=1}^{n_2} \chi_{il}, \tag{B.51}$$

$$\Lambda \phi_j = \xi' \sum_{m=1}^{n_1} k_{mj} \chi_{mj}. \tag{B.52}$$

Here we have, as in Section 8, put $c_{ij} = 1$.

Using eqn (B.50) to substitute for χ_{ij} in eqns (B.51) and (B.52), we have

$$\Lambda^2 v_i + \xi \sum_{l=1}^{n_2} v_{il}^*(v_i + \phi_l) = 0, \tag{B.53}$$

$$\Lambda^2 \phi_j + \xi' \sum_{m=1}^{n_1} k_{mj} v_{mj}^*(v_m + \phi_j) = 0. \tag{B.54}$$

Using the equilibrium expressions given by eqn (14.43), we can reduce this pair of equations to

$$(\Lambda^2 + \xi b)v_i + \xi \sum_{l=1}^{n_2} v_{il}^* \phi_l = 0, \tag{B.55}$$

$$\xi' \sum_{m=1}^{n_1} k_{mj} v_{mj}^* v_m + (\Lambda^2 + \xi' b)\phi_j = 0. \tag{B.56}$$

Using eqn (B.56) to substitute for ϕ_l in eqn (B.55) leads us finally to a set of n_1 equations for the perturbations v_i:

$$\sum_{j=1}^{n_1} [A_{ij} - F(\Lambda)\delta_{ij}]v_j = 0. \tag{B.57}$$

Here A is the $n_1 \times n_1$ matrix with elements

$$A_{ij} \equiv \xi \xi' \sum_{l=1}^{n_2} v_{il}^* v_{jl}^* k_{jl}, \tag{B.58}$$

and $F(\Lambda)$ is a fourth-order polynomial in the stability-determining quantities Λ

$$F(\Lambda) = (\Lambda^2 + \xi b)(\Lambda^2 + \xi' b). \tag{B.59}$$

That is, $F(\Lambda)$ is a quadratic in Λ^2:

$$F(\Lambda) = (\Lambda^2)^2 + br(\Lambda^2) + \xi \xi' b^2. \tag{B.60}$$

Here we have used $\xi + \xi' = r$.

The expression (B.57) is reminiscent of eqn (B.19) of Appendix B.2. We again observe that, if λ_i are the eigenvalues of the matrix A of eqn (B.58) (with $i = 1, 2, \ldots, n_1$), then Λ^2 are given from the quadratic equations $F(\Lambda) = \lambda_i$, or

$$(\Lambda^2)^2 + br(\Lambda^2) + (\xi \xi' b^2 - \lambda_i) = 0. \tag{B.61}$$

Unless both roots of all n_1 such quadratic equations for Λ^2 are real and negative, there will be at least one stability-determining rate Λ with positive real part, implying that the interior fixed point is unstable. But if *all* Λ^2 are real and negative, then the interior fixed point will be (locally) neutrally stable. This latter event requires $b^2 r^2 > 4(\xi \xi' b^2 - \lambda_i) > 0$, for all i. From the definition of the matrix elements of A, eqn (B.58), we can rescale $\lambda_i = \xi \xi' b^2 \lambda_i'$ (rescaling the v_{ij}^* to v_{ij}^*/b), so that the constant term in eqn (B.61) reads as $\xi \xi' b^2 (1 - \lambda_i')$. For many matrices k_{ij}, the equilibrium values v_{ij}^* will indeed lead to $\lambda_i' < 1$, so that the conditions for a (locally) neutrally stable equilibrium (namely, $r^2 > 4\xi(r - \xi)(1 - \lambda_i') > 0$) are satisfied, provided ξ or $(r - \xi)$ is sufficiently small relative to r.

REFERENCES

Adorini, L., Appella, E., Doria, G. and Nagy, Z. A. (1988). Mechanisms influencing the immunodominance of T-cell determinants. *J. Exp. Med.*, **168**, 2091–104.

Aebischer, T., Moskophidis, D., Rohrer, U. H., Zinkernagel, R. M. and Hengartner, H. (1991). *In vitro* selection of lymphocytic choriomeningitis virus escape mutants by cytotoxic T lymphocytes. *Proc. Natl. Acad. Sci. USA*, **88**, 11047–51.

Agur, Z. (1989). Clinical trials of zidovudine in HIV infection. *Lancet*, **2**, 1400.

Agur, Z., Abiri, D. and Van der Ploeg, L. H. (1989). Ordered appearance of antigenic variants of African trypanosomes explained in a mathematical model based on a stochastic switch process and immune-selection against putative switch intermediates. *Proc. Natl. Acad. Sci. USA*, **86**, 9626–30.

Agur, Z., Arnon, R., Sandak, B. and Schechter, B. (1991). Zidovudine toxicity to murine bone marrow may be affected by the exact frequency of drug administration. *Exp. Hematol.*, **19**, 364–8.

Albert, J., Abrahamsson, B., Nagy, K., Aurelius, E., Gaines, H., Nystrom, G. and Fenyo, E. M. (1990). Rapid development of isolate-specific neutralizing antibodies after primary HIV-1 infection and consequent emergence of virus variants which resist neutralization by autologous sera. *AIDS*, **4**, 107–12.

Anderson, R. M. and May, R. M. (1979*a*). Population biology of infectious diseases: Part 1. *Nature*, **280**, 361–7.

Anderson, R. M. and May, R. M. (1979*b*). Population biology of infectious diseases. II. *Nature*, **280**, 455.

Anderson, R. M., May, R. M. and Gupta, S. (1989). Non-linear phenomena in host–parasite interactions. *Parasitology*, **99**(Suppl.), S59–79.

Anderson, R. M. and May, R. M. (1991). *Infectious diseases of humans*, Oxford University Press.

Antia, R. N. and Koella, J. C. (1994). A model of non-specific immunity, *J. Theor. Biol.*, **168**, 141–50.

Arias, I., Boyer, J. L., Fausto, N., *et al.* (2000). *The liver: biology and pathobiology*, 4th edn, Raven Press.

Arnaout, R. A., Lloyd, A. L., O'Brien, T. R., Goedert, J. J., Leonard, J. M. and Nowak, M. A. (1999). A simple relationship between viral load and survival time in HIV-1 infection. *Proc. Natl. Acad. Sci. USA*, **96**, 11549–53.

Asjo, B., Sharma U. K., Morfeldt-Manson L., Magnusson, A., Barkhem, T., Albert, J., Olausson E., Von Gegerfelt, A., Lind, B., Biberfeld, P., *et al.* (1990). Naturally occurring HIV-1 isolates with differences in replicative capacity are distinguished by *in situ* hybridization of infected cells. *AIDS Res. Hum. Retrovir.*, **6**, 1177–82.

Baire, M., Dittmar, M. T., Cichutek, K. and Kurth, R. (1980). Development *in vivo* of genetic variability of SIV. *Proc. Natl. Acad. Sci. USA*, **88**, 8126–30.

Balfe, P., Simmonds, P., Ludlam, C. A., Bishop, J. O. and Brown, A. J. L. (1990). Concurrent evolution of human-immunodeficiency-virus type-1 in patients infected from the same source—rate of sequence change and low-frequency of inactivating mutations. *J. Virol.*, **64**, 6221–33.

Bangham, C. R. (1993). Human T-cell leukaemia virus type I and neurological disease. *Curr. Opin. Neurobiol.*, **3**, 773–8.

Bangham, C. R. M., Kermode, A. G., Hall, S. E. and Daenke, S. (1996). The cytotoxic T-lymphocyte response to HTLV-I: the main determinant of disease? *Semin. Virol.*, **7**, 41–8.

Bangham, C. R., Hall, S. E., Jeffery, K. J., Vine, A. M., Witkover, A., Nowak, M. A., Wodarz, D., Usuku, K. and Osame, M. (1999). Genetic control and dynamics of the cellular immune response to the human T-cell leukaemia virus, HTLV-I. *Philos. Trans. R. Soc. Lond. B Biol. Sci.*, **354**, 691–700.

Barre-Sinoussi, F., Chermann, J. C., Rey, F., Nugeyre, M. T., Charmaret, S., Gruest, J., Dauguet, C., Axler-Blin, C., Vezinet-Brun, F., Rouzioux, C. and Rozenbaum, W. (1983). Isolation of a T-lymphotropic retrovirus from a patient at risk for acquired immune deficiency syndrome (AIDS). *Science*, **220**, 868–71.

Bell, G. I. (1970*a*). Mathematical model of clonal selection and antibody production. *Nature*, **228**, 739–44.

Bell, G. I. (1970*b*). Mathematical model of clonal selection and antibody production. *J. Theor. Biol.*, **29**, 191–232.

Bell, G. I. (1971*a*). Mathematical model of clonal selection and antibody production. II. *J. Theor. Biol.*, **33**, 339–78.

Bell, G. I. (1971*b*). Mathematical model of clonal selection and antibody production. III. The cellular basis of immunological paralysis. *J. Theor. Biol.*, **33**, 378–98.

Bertoletti, A., Sette, A., Chisari, F. V., Penna, A., Levrero, M., Decarli, M., Fiaccadori, F. and Ferrari, C. (1994). Natural variants of cytotoxic epitopes are T cell receptor antagonists for antiviral cytotoxic T cells. *Nature*, **369**, 407–110.

Biebricher, C. K., Eigen, M. and Gardiner, W. C. (1985). Kinetics of RNA replication: competition and selection among self-replicating RNA species. *Biochemistry-US*, **24**, 6550–60.

Biebricher, C. K., Eigen, M. and Luce, R. (1986). Template-free RNA synthesis by Q-beta replicase. *Nature*, **321**, 89–92.

Biggar, R. J. (1990). AIDS incubation in 1891 HIV seroconverters from different exposure Groups. *AIDS*, **4**, 1059–10663.

Bittner, B., Bonhoeffer, S. and Nowak, M. A. (1997). Virus load and antigenic diversity. *Bull. Math. Biol.*, **59**, 881–96.

Bonhoeffer, S. and Nowak, M. A. (1997). Pre-existence and emergence of drug resistance in HIV-1 infection. *Proc. Roy. Soc. Lond. B.*, **264**, 631–7.

Bonhoeffer, S., Holmes, E. C. and Nowak, M. A. (1995). Causes of HIV diversity. *Nature*, **376**, 125.

Bonhoeffer, S., May, R. M., Shaw, G. M. and Nowak, M. A. (1997*a*). Virus dynamics and drug therapy. *Proc. Natl. Acad. Sci. USA*, **94**, 6971–6.

Bonhoeffer, S., Coffin, J. M. and Nowak, M. A. (1997*b*). Human immunodeficiency virus drug therapy and virus load. *J. Virol.*, **71**, 3275–8

Borghans, J. A., de Boer, R. J. and Segel, L. A. (1996). Extending the quasi-steady state approximation by changing variables. *Bull. Math. Biol.*, **58**, 43–63.

Borrow, P., Lewicki, H., Hahn, B. H., Shaw, G. M. and Oldstone, M. B. A. (1994). Virus specific CD8+ cytotoxic T lymphocyte activity associated with control of viremia in primary human immunodeficiency virus type-1 infection. *J. Virol.*, **68**, 6103–10.

Borrow, P., Lewicki, H., Wei, X. P., *et al.* (1997). Antiviral pressure by HIV-1 specific cytotoxic T lymphocytes (CTLs) during primary infection demonstrated by rapid selection of CTL escape virus. *Nature Med.*, **3**, 205–11.

Boucher, C. A., Tersmette, M., Lange, J. M., Kellam, P., de Goede, R. E., Mulder, J. W., *et al.* (1990). Zidovudine sensitivity of human immunodeficiency viruses from high-risk, symptom-free individuals during therapy. *The Lancet*, **336**, 585–90.

Boucher, C. A., Lange, J. M., Miedema, F. F., Weverling, G. J., Koot, M., Mulder, J. W., *et al.* (1992*a*). HIV-1 biological phenotype and the development of zidovudine resistance in relation to disease progression in asymptomatic individuals during treatment. *AIDS*, **6**, 1259–64.

Boucher, C. A., O'Sullivan, E., Mulder, J. W., Ramautarsing, C., Kellam, P., *et al.* (1992*b*). Ordered appearance of zidovudine resistance mutations during treatment of 18 human immunodeficiency virus-positive subjects. *J. Infect. Dis.*, **165**, 105–10.

Bukrinsky, M. I., Stanwick, T. L., Dempsey, M. P. and Stevenson, M. (1991). Quiescent lymphocytes-T as an inducible virus reservoir in HIV-1 infection. *Science*, **254**, 423–7.

Burk, R. D., Hwang, L. Y., Ho, G. Y., Shafritz, D. A. and Beasley, R. P. (1994). Outcome of perinatal hepatitis B virus exposure is dependent on maternal virus load. *J. Infect. Dis.*, **170**, 1418–23.

Burns, D. P. W. and Desrosiers, R. C. (1991). Selection of genetic variants of SIV in persistently infected rhesus monkeys. *J. Virol.*, **65**, 1843–54.

Buseyne, F., Blanche, S., Schmitt, D., Griscelli, C. and Riviere Y. (1993). Detection of HIV-specific cell-mediated cytotoxicity in the peripheral blood from infected children. *J. Immunol.*, **150**, 3569–81.

Callaway, D. S., Ribeiro, R. M. and Nowak, M. A. (1999). Virus phenotype switching and disease progression in HIV-1 infection. *Proc. R. Soc. Lond. B Biol. Sci.*, **266**, 2523–30.

Campos-Lima, P., Gavioli, R., Zhang, Q., Wallace, L. E., Dolcetti, R., Rowe, M., Rickinson, A. B. and Masucci, M. G. (1993). HLA-A11 epitope loss isolates of Epstein–Barr virus from highly A11+ population. *Science*, **260**, 98–100.

Carpenter, S., Evans, L. H., Sevoian, M. and Chesebro, B. (1990). *In vivo* and *in vitro* selection of equine infectious anemia virus variants. In: *Applied virology research, Vol. 2, Virus variability, epidemiology, and control* (Kurstak, E., Marusyk, R. G., Murphy, F. A. and van Regenmortel, H. V., ed.) Plenum, 99–115.

Carrington, M., Nelson, G. W., Martin, M. P., Kissner, T., Vlahov, D., Goedert, J. J., Kaslow, R., Buchbinder, S., Hoots, K. and O'Brien, S. J. (1999). HLA and HIV-1: heterozygote advantage and B*35-Cw*04 disadvantage. *Science*, **283**, 1748–52.

Celada, F. and Seiden, P. E. (1992). A computer model of cellular interactions in the immune system. *Immunol. Today*, **13**, 56–62.

Celada, F. and Seiden, P. E. (1996). Affinity maturation and hypermutation in a simulation of the humoral immune response. *Eur. J. Immunol.*, **26**, 1350–8.

Cerny, A. and Chisari, F. V. (1999). Pathogenesis of chronic hepatitis C: immunological features of hepatic injury and viral persistence. *Hepatology*, **30**, 595–601.

Chisari, F. V. (1997). Cytotoxic T cells and viral hepatitis. *J. Clin. Invest.*, **99**, 1472–7.

Chisari, F. V. and Ferrari, C. (1995). Hepatitis-B virus immunopathogenesis. *Annu. Rev. Immunol.*, **13**, 29–60.

Chiu, I. M., Yaniv, A., Dahlberg, J. E., Gazit, A., Skuntz, S. F., Tronick, S. R. and Aaronson, S. A. (1985). Nucleotide sequence evidence for relationship of AIDS retrovirus to lentiviruses. *Nature*, **317**, 366–8.

Chun T. W. and Fauci A. S. (1999). Latent reservoirs of HIV: obstacles to the eradication of virus. *Proc. Natl. Acad. Sci. USA*, **96**, 10958–61.

Chun, T. W., Carruth, L., Finzi, D., Shen, X., DiGiuseppe, J. A., Taylor, H., Hermankova, M., Chadwick, K., Quinn, T. C., Kuo, Y. H., Brookmeyer, R., Zeiger, M. A., Barditch-Crovo, P. and Siliciano, R. F. (1997*a*). Quantification of latent tissue reservoirs and total body viral load in HIV-1 infection. *Nature*, **387**, 183–8.

Chun, T. W., Stuyver, L., Mizell, S. B., Ehler, L. A., Mican, J. A. M., Baseler, M., Lloyd, A. L., Nowak, M. A. and Fauci, A. S. (1997*b*). Presence of an inducible HIV-1 latent reservoir during highly active antiretroviral therapy. *Proc. Natl. Acad. Sci. USA*, **94**, 13193–7.

Chun, T. W., Davey, Jr., R. T., Engel, D., Lane, H. C. and Fauci, A. S. (1999). Re-emergence of HIV after stopping therapy. *Nature*, **401**, 874–5.

Clavel, F., Guyader M., Guetard, D., Salle, M., Montagnier, L. and Alizon, M. (1986*a*). Molecular cloning and polymorphism of the human immune deficiency virus type 2. *Nature*, **324**, 691–5.

Clavel, F., Guetard, D., Brun-Vezinet, F., Chamaret, S., Rey, M. A., Santos-Ferreira, M. O., Laurent, A. G., Dauguet, C., Katlama, C., Rouzioux, C., *et al.* (1986*b*). Isolation of a new human retrovirus from West African patients with AIDS. *Science*, **233**, 343–6.

Clements, J. E., Pedersen, F. S., Narayan, O. and Haseltine, W. A. (1980). Genomic changes associated with antigenic variation of visna virus during persistant infection. *Proc. Natl. Acad. Sci. USA*, **77**, 4454–8.

Cocchi, F., Devico, A. L., Garzinodemo, A., Arya, S. K., Gallo, R. C. and Lusso, P. (1995). Identification of RANTES, Mip-1-alpha and MIP-1-beta as the major HIV suppressive factors produced by CD8+ T-cells. *Science*, **270**, 1811–15.

Coffin, J. M. (1995). HIV population dynamics *in vivo*: implications for genetic variation, pathogenesis and therapy. *Science*, **267**, 483–9.

Coffin, J. M., Hughes, S. H. and Varmus H. E. (ed.) (1997). *Retroviruses*, Cold Spring Harbor Laboratory Press.

Condra, J. H., Schleif, W. A., Blahy, O. M., *et al.* (1995). *In-vivo* emergence of HIV-1 variants resistant to multiple protease inhibitors. *Nature*, **374**, 569–71.

Connor, R. I. and Ho, D. D. (1994). Human immunodeficiency virus type 1 variants with increased replicative capacity develop during the asymptomatic stage before disease progression. *J. Virol.*, **68**, 4400–8.

Connor, R. I., Mohri, H., Cao, Y. Z. and Ho, D. D. (1993). Increased viral burden and cytopathicity correlate temporally with CD4+ T-lymphocyte decline and clinical progression in human immunodeficiency virus type-1 infected individuals. *J. Virol.*, **67**, 1772–7.

Connor, R. I., Sheridan, K. E., Ceradini, D., Choe, S. and Landau, N. R. (1997). Change in coreceptor use correlates with disease progression in HIV-1 infected individuals. *J. Exp. Med.*, **185**, 621–8.

Daenke, S., Kermode, A. G., Hall, S. E., Taylor, G., Weber, J., Nightingale, S. and Bangham, C. R. (1996). High activated and memory cytotoxic T-cell responses to HTLV-1 in healthy carriers and patients with tropical spastic paraparesis. *Virology*, **217**, 139–46.

Dalgleish, A. G., Beverley, P. C., Clapham, P. R., Crawford, D. H., Greaves, M. F. and Weiss, R. A. (1984). The CD4 (T4) antigen is an essential component of the receptor for the AIDS retrovirus. *Nature*, **312**, 760–3.

Dean, M., Carrington, M., Winkler, C., *et al.* (1996). Genetic restriction of HIV-1 infection and progression to AIDS by a deletion allele of the CKR5 structural gene. *Science*, **273**, 1856–62.

De Boer, R. J. and Boerlijst, M. C. B. (1994). Diversity and virulence thresholds in AIDS. *Proc. Nat. Acad. Sci. USA*, **91**, 544–8.

De Boer, R. J. and Boucher, C. A. B. (1996). Anti-CD4 therapy of AIDS suggested by mathematical models. *Proc. R. Soc. Lond. B Biol. Sci.*, **263**, 899–905.

De Boer, R. J. and Noest, A. J. (1998). T cell renewal rates, telomerase and telomere length shortening. *J. Immunol.*, **160**, 5832–7.

De Boer, R. J. and Perelson, A. S. (1995). Towards a general function describing T cell proliferation. *J. Theor. Biol.*, **175**, 567–76.

De Boer, R. J. and Perelson, A. S. (1998). Target cell limited and immune control models of HIV infection: a comparison. *J. Theor. Biol.*, **190**, 201–14.

De Boer, R. J., Segel, L. A. and Perelson A. S. (1992). Pattern formation in one- and two-dimensional shape-space models of the immune system. *J. Theor. Biol.*, **155**, 295–33.

De Boer, R. J., Boucher, C. A. and Perelson, A. S. (1998). Target cell availability and the successful suppression of HIV by hydroxyurea and didanosine. *AIDS*, **12**, 1567–70.

De Jong, M. D., Veenstra, J., Stilianakis, N. I., Schuurman, R., Lange, J. M. A., de Boer, R. J. and Boucher, C. A. B. (1996). Host–parasite dynamics and outgrowth of virus containing a single K70R amino acid change in reverse transcriptase are responsible for the loss of HIV-1 RNA load suppression by zidovudine. *Proc. Nat. Acad. Sci. USA*, **93**, 5501–6.

Delassus, S., Cheynier, R. and Wain-Hobson, S. (1992). Nonhomogeneous distribution of human immunodeficiency virus type 1 proviruses in the spleen. *J. Virol.*, **66**, 5642–5.

Delwart, E. L., Sheppard, H. W., Walker, B. D., Goudsmit, J. and Mullins, J. I. (1994). Human immunodeficiency virus type 1 evolution *in vivo* tracked by DNA heteroduplex mobility assays. *J. Virol.*, **68**, 6672–83.

Delwart, E. L., Pan, H., Sheppard, H. W., Wolpert, D., Neumann, A. U., Korber, B. and Mullins, J. I. (1997). Divergent patterns of progression to AIDS after infection from the same source: human immunodeficiency virus type 1 evolution and antiviral responses. *J. Virol.*, **71**, 4284–95.

Diamond, J. M. (1997). *Guns, germs and steel: the fates of human societies*, W.W. Norton & Company.

Diekmann, O. and Heesterbeek, H. (2000). *Mathematical epidemiology of infectious diseases*, John Wiley & Sons.

Dienstag, J. L., Perrillo, R. P., Schiff, E. R., Bartholomew, M., Vicary, C. and Rubin, M. (1994). Double-blind, randomized, 3-month, dose-ranging trial of lamivudine for chronic hepatitis B infection. *Hepatology*, **20**, 199–205.

Dienstag, J. L., Perillo, R. P., Schiff, E. R., Bartholomew, M., Vicary, C. and Rubin, M. (1995). A preliminary trial of lamivudine for chronic hepatitis B infection. *New Eng. J. Med.*, **333**, 1657–61.

Dimmock, N. J. and Primrose, S. B. (1994). *Introduction to modern virology*. 4th edn. Basic Microbiology Series, Blackwell.

Dittmar, M. T., McKnight, A., Simmons, G., Clapham, P. R., Weiss, R. A. and Simmonds, P. (1997). HIV tropism and coreceptor use. *Nature*, **385**, 495–6.

Domingo, E., Sabo, D., Taniguchi, T. and Weissmann, C. (1978). Nucleotide sequence heterogeneity of an RNA phage population. *Cell*, **13**, 735–44.

Domingo, E., Davila, M. and Ortin, J. (1980). Nucleotide sequence heterogeneity of the RNA from a natural population of foot-and-mouth disease virus. *Gene*, **11**, 333–46.

Domingo, E., Diez, J., Martinez, M. A., Hernandez, J., Holguin, A., Borrego, B. and Matteu, M. G. (1993). New observations on antigenic diversification of RNA viruses. Antigenic variation is not dependent on immune selection. *J. Gen. Virol.*, **74**, 2039–45.

Doong, S. L., Tsai, C. H., Schinazi, R. F., Liotta, D. C. and Cheng, Y. C. (1991). Inhibition of the replication of hepatitis B virus *in vitro* by 2′3′-dideoxy-3′-thiacytidine and related analogues. *Proc. Natl. Acad. Sci. USA*, **88**, 8495–9.

Dragic, T., Litwin, V., Allaway, G. P., *et al.* (1996). HIV-1 entry into CD4(+) cells is mediated by the chemokine receptor CC-CKR-5. *Nature*, **381**, 667–73.

Duarte, E. A., Novella, I. S., Weaver, S. C., Domingo, E. Wain-Hobson, S., Clarke, D. K., Moya, A., Elena, S. F., de la Torre, J. C. and Holland, J. J. (1994). RNA virus quasispecies: significance for viral disease and epidemiology. *Infect. Agents Dis.*, **3**, 201–14.

Edelstein-Keshet, L. (1988). *Mathematical models in biology*, McGraw-Hill.

Eigen, M. (1971). Self-organization of matter and the evolution of biological macromolecules. *Naturwissenschaften*, **58**, 465–523.

Eigen, M. and Schuster, P. (1977). The hypercycle. A principle of natural self-organisation. Part A: Emergence of the hypercycle. *Naturwissenschaften*, **64**, 541–65.

Eigen, M., McCaskill, J. and Schuster, P. (1989). The molecular quasispecies. *Adv. Chem. Phys.*, **75**, 149–263.

Ellis, T. M., Wilcox, G. E. and Robinson, W. F. (1987). Antigenic variation of CAEV during persistent infection of goats. *J. Gen. Virol.*, **68**, 3145–52.

Eron, J. J., Benoit, S. L., Jemsek, J., MacArthur, R. D., *et al.* (1995). Treatment with lamivudine, zidovudine, or both in HIV-positive patients with 200–500 CD4+ cells per cubic millimeter. *N. Engl. J. Med.*, **333**, 1662–9

Essunger, P. and Perelson, A. S. (1994). Modelling HIV-infection of CD4(+) T cell subpopulations. *J. Theor. Biol.*, **170**, 367–91.

Fabrizi, F., Martin, P., Dixit, V., Brezina, M., Cole, M. J., Gerosa, S., Vinson, S., Mousa, M. and Gitnick, G. (1998). Quantitative assessment of HCV load in chronic hemodialysis patients: a cross-sectional survey. *Nephron.*, **80**, 428–33.

Fauci, A. S. (1996). Host factors and the pathogenesis of HIV-induced disease. *Nature*, **384**, 529–34.

Feinberg, M. B., Jarrett, R. F., Aldovini, A., Gallo, R. C. and Wong-Staal, F. (1986). HTLV-III expression and production involve complex regulation at the levels of splicing and translation of viral RNA. *Cell*, **46**, 807–17.

Feldman, M. W. and Karlin, S. (1989). *Mathematical evolutionary theory*, Princeton University Press.

Fenouillet, E., Blanes, N., Benjouas A. and Gluckman, J. C. (1995). Anti-V3 antibody reactivity correlates with clinical stage of HIV-1 infection and with serum neutralising activity. *Clinic. Experim. Immunol.*, **99**, 419–24.

Fenyo, E. M. (1994). Antigenic variation of primate lentiviruses in humans and experimentally infected macaques. *Immunol. Rev.*, **140**, 131–46.

Ferbas, J., Daar, E. S., Grovit-Ferbas, K., Lech, W. J., Detels, R., Giorgi, J. V. and Kaplan, A. H. (1996). Rapid evolution of human immunodeficiency virus strains with increased replicative capacity during the seronegative window of primary infection. *J. Virol.*, **70**, 7285–9.

Finzi, D., Hermankova, M., Pierson, T., Carruth, L. M., Buck, C., Chaisson, R. E., Quinn, T. C., Chadwick, K., Margolick, J., Brookmeyer, R., Gallant, J. Markowitz, M., Ho, D. D., Richman, D. D. and Siliciano, R. F. (1997). Identification of a reservoir of HIV-1 in patients on highly active antiretroviral therapy. *Science*, **278**, 1295–300.

Finzi, D., Blankson, J., Siliciano, J. D., Margolick, J. B., Chadwick, K., Pierson, T., Smith, K., Lisziewicz, J., Lori, F., Flexner, C., Quinn, T. C., Chaisson, R. E., Rosenberg, E., Walker, B., Gange, S., Gallant, J. and Siliciano, R. F. (1999). Latent infection of CD4+ T cells provides a mechanism for lifelong persistence of HIV-1, even in patients on effective combination therapy. *Nat. Med.*, **5**, 512–17.

Fisher, A. G., Collalti, E., Ratner, L., Gallo, R. C. and Wong-Staal, F. (1985). A molecular clone of HTLV-III with biological activity. *Nature*, **316**, 262–5.

Fisher, A. G., Ensoli, B., Looney, D., *et al.* (1988). Biologically diverse molecular variants within a single HIV-1 isolate. *Nature*, **334**, 444–7.

Fishman, M. A. and Perelson, A. S. (1995). Lymphocyte memory and affinity selection. *J. Theor. Biol.*, **173**, 241–62.

Fong, T. L., Di Bisceglie, A. M., Biswas, R., Waggoner, J. G., Wilson, L., Claggett, J. and Hoofnagle, J. H. (1994). High levels of viral replication during acute hepatitis B infection predict progression to chronicity. *J. Med. Virol.*, **43**, 155–8.

Fontana, W. and Schuster, P. (1987). A computer model of evolutionary optimization. *Biophys. Chem.*, **26**, 123–47.

Fontana, W. and Schuster, P. (1998). Continuity in evolution: on the nature of transitions. *Science*, **280**, 1451–5.

Fontana, W., Schnabl, W. and Schuster, P. (1989). Physical aspects of evolutionary optimization and adaptation. *Phys. Rev. A*, **40**, 3301–21.

Frost, S. D. W. and McLean, A. R. (1994). Quasispecies dynamics and the emergence of drug resistance during zidovudine therapy of HIV infection. *AIDS*, **8**, 323–32.

Gallo, R. C. and Lusso, P. (1997). Chemokines and HIV infection. *Curr. Op. Inf. Dis.*, **10**, 12–17.

Gallo, R. C., Sarin, P. S., Gelmann, E. P., Robert-Guroff, M., Richardson, E., Kalyanaraman, V. S., Mann, D., Sidhu, G. D., Stahl, R. E., Zolla-Pazner, S., Liebowitch, J. and Popovic, M. (1983). Isolation of human T-cell leukemia virus in acquired immune deficiency syndrome (AIDS). *Science*, **220**, 865–7.

Ganeshan, S., Dickover, R. E., Korber, B. T., Bryson, Y. J. and Wolinsky, S. M. (1997). Human immunodeficiency virus type 1 genetic evolution in children with different rates of development of disease. *J. Virol.*, **71**, 663–77.

Gelmann, E. P., Popvic, M., Blayney, D., Masur, H., Sidhu, G., Stahl, R. E. and Gallo, R. C. (1983). Proviral DNA of a retrovirus, human T-cell leukemia virus, in two patients with AIDS. *Science*, **220**, 862–5.

Gotch, F. M., Nixon, D. F., Alp, N., McMichael, A. J. and Borysiewicz, L. K. (1990). High-frequency of memory and effector Gag specific cytotoxic lymphocytes-T in HIV seropositive individuals. *Int. Immunol.*, **2**, 707.

Goudsmit, J. (1988). Immunodominant B-cell epitopes of the HIV-1 envelope. *AIDS*, **2**(Suppl. 1), S41–5.

Goudsmit, J. (1995). The role of viral diversity in HIV pathogenesis. *J. Acquir. Immune Defic. Syndr. Hum. Retrovirol.*, **10**(Suppl.), S15–19.

Goudsmit, J. (1997). *Viral sex: the nature of AIDS*, Oxford University Press.

Goudsmit, J., Back, N. K. and Nara, P. L. (1991). Genomic diversity and antigenic variation of HIV-1: links between pathogenesis, epidemiology and vaccine development. *FASEB J.*, **5**, 2427–36.

Goudsmit, J., De Ronde, A., Ho, D. D. and Perelson, A. S. (1996). Human immunodeficiency virus fitness *in vivo*: calculations based on a single zidovudine resistance mutation at codon 215 of reverse transcriptase. *J. Virol.*, **70**, 5662–4.

Goulder, P. J. R., Phillips, R. E., Colbert, R. A., *et al.* (1997). Late escape from an immunodominant cytotoxic T-lymphocyte response associated with progression to AIDS. *Nat. Med.*, **3**, 212–17.

Grossman, Z., Polis, M., Feinberg, M. B., Levi, I., Jankelevich, S., Yarchoan, R., Boon, J., de Wolf, F., Lange, J. M., Goudsmit, J., Dimitrov, D. S. and Paul, W. E. (1999). Ongoing HIV dissemination during HAART. *Nat. Med.*, **5**, 1099–104.

Guidotti, L. G. and Chisari, F. V. (1996). To kill or to cure: options in host-defense against viral infection. *Curr. Opin. Immunol.*, **8**, 478–83.

Guidotti, L. G., Ando, K., Hobbs, M. V., Ishikawa, T., Runkel, L., Schreiber, R. D. and Chisari, F. V. (1994a). Cytotoxic T-lymphocytes inhibit hepatitis-B virus gene expression by a noncytolytic mechanism in transgenic mice. *Proc. Natl. Acad. Sci. USA*, **91**, 3764–8.

Guidotti, L. G., Guilhot, S. and Chisari, F. V. (1994*b*). Interleukin-2 and alpha/beta interferon down-regulate hepatitis-B virus gene-expression *in-vivo* by tumor necrosis factor-dependent and factor-independent pathways. *J. Virol.*, **68**, 1265–70.

Guidotti, L. G., Ishikawa, T., Hobbs, M. V., Matzke, B., Schreiber, R. and Chisari, F. V. (1996*b*). Intracellular inactivation of the hepatitis-B virus by cytotoxic T lymphocytes. *Immunity*, **4**, 25–36.

Guidotti, L. G., Rochford, R., Chung, J., Shapiro, M., Purcell, R. and Chisari F. V. (1999). Viral clearance without destruction of infected cells during acute HBV infection. *Science*, **284**, 825–9.

Gupta, S. and Anderson, R. M. (1999). Population structure of pathogens: the role of immune selection. *Parasitol. Today*, **15**, 497–501.

Gupta, S., Trenholme, K., Anderson, R. M. and Day, K. P. (1994). Antigenic diversity and the transmission dynamics of *Plasmodium falciparum*. *Science*, **263**, 961–3.

Gupta, S., Maiden, M. C. J., Feavers, I. M., Nee, S., May, R. M. and Anderson, R. M. (1996). The maintenance of strain structure in populations of recombining infectious agents. *Nat. Med.*, **2**, 437–42.

Gupta, S., Ferguson, N. and Anderson, R. (1998). Chaos, persistence and evolution of strain structure in antigenically diverse infectious agents. *Science*, **280**, 912–15.

Haase, A. T. (1986). Pathogenesis of lentivirus infections. *Nature*, **322**, 130–6.

Haase, A. T., Henry, K., Zupancic, M., Sedgewick, G., Faust, R. A., Melroe, H., Cavert, W., Gebhard K., Staskus, K., Zhang, Z. Q., Dailey, P. J., Balfour, Jr., H. H., Erice, A. and Perelson, A. S. (1996). Quantitative image analysis of HIV-1 infection in lymphoid tissue. *Science*, **274**, 985–9.

Hahn, B. H., Shaw, G. M., Arya, S. K., Popovic, M., Gallo, R. C. and Wong-Staal, F. (1984). Molecular cloning and characterization of the HTLV-III virus associated with AIDS. *Nature*, **312**, 166–9.

Hahn, B. H., Shaw, G. M., Taylor, M. E., Redfield, R. R., Markham, P. D., Salahuddin, S. Z., Wong-Staal, F., Gallo, R. C., Parks, E. S. and Parks W. P. (1986). Genetic variation in HTLV-III/LAV over time in patients with AIDS or at risk for AIDS. *Science*, **232**, 1548–53.

Haynes, B. F., Pantaleo, G. and Fauci, A. S. (1996). Towards an understanding of the correlates of protective immunity to HIV infection. *Science*, **271**, 324–8.

Herz, A. V. M., Bonhoeffer, S., Anderson, R. M., May, R. M. and Nowak, M. A. (1996). Viral dynamics *in vivo*: limitations on estimates on intracellular delay and virus decay. *Proc. Natl. Acad. Sci. USA*, **93**, 7247–51.

Hetzel, C. and Anderson, R. M. (1996). The within-host cellular dynamics of bloodstage malaria: theoretical and experimental studies. *Parasitology*, **113**, 25–38.

Hill, A. V. S. (1997). MHC polymorphism and susceptibility to intracellular infections in humans. In *Host response to intracellular pathogens* (Kaufmann, S. H. E. and Landes, R. G., ed.), pp. 47–59. Lippincott, Williams & Wilkins.

Hlavacek, W. S., Wofsy, C. and Perelson, A. S. (1999). Dissociation of HIV-1 from follicular dendritic cells during HAART: mathematical analysis. *Proc. Natl. Acad. Sci. USA*, **96**, 14681–6.

Ho, D. D., Toyoshima, T., Mo, H., Kempf, D. J., Norbeck, D., Chen, C. M., *et al.* (1994). Characterization of human immunodeficiency virus type 1 variants with increased resistance to a C2-symmetric protease inhibitor. *J. Virol.*, **68**, 2016–20.

Ho, D. D., Neumann, A. U., Perelson, A. S., *et al.* (1995). Rapid turnover of plasma virions and CD4 lymphocytes in HIV-1 infection. *Nature*, **373**, 123–6.

Hockett, R. D., Kilby, J. M., Derdeyn, C. A., Saag, M. S., Sillers, M., Squires, K., Chiz, S., Nowak, M. A., Shaw, G. M. and Bucy, R. P. (1999). Constant mean viral copy number per infected cell in tissues regardless of high, low, or undetectable plasma HIV RNA. *J. Exp. Med.*, **189**, 1545–54.

Hofbauer, J. and Sigmund, K. (1998). *The theory of evolution and dynamical systems*, Cambridge University Press.

Holland, J., Spindler, K., Horodyski, F., Grabeu, E., Nichol, S. and Vandepol, S. (1982). Rapid evolution of RNA genomes. *Science*, **215**, 1577–85.

Holmes, E. C., Zhang, L. Q., Simmonds, P., Ludlam, C. A. and Leigh Brown, A. J. (1992). Convergent and divergent sequence evolution in the surface envelope glycoprotein of HIV-1 within a single infected patient. *Proc. Natl. Acad. Sci. USA*, **89**, 4835–9.

Jacobson, S., Shida, H., McFarlin, D. E., Fauci, A. S. and Koenig S. (1990). Circulating CD8+ cytotoxic lymphocytes-T specific for HTLV-I PX in patients with HTLV-I associated neurological disease. *Nature*, **348**, 245–8.

Janeway, C., *et al.* (1999). *Immunobiology: the immune system in health and disease*, Garland.

Jeffery, K. J., Usuku, K., Hall, S. E., Matsumoto, W., Taylor, G. P., Procter, J., Bunce, M., Ogg, G. S., Welsh, K.I., Weber, J. N., Lloyd, A. L., Nowak, M. A., Nagai, M., Kodama, D., Izumo, S., Osame, M. and Bangham, C. R. (1999). HLA alleles determine human T-lymphotropic virus-I (HTLV-I) proviral load and the risk of HTLV-I-associated myelopathy. *Proc. Natl. Acad. Sci. USA*, **96**, 3848–53.

Jin, X., Bauer, D. E., Tuttleton, S. E., Lewin, S., Gettie, A., Blanchard, J., Irwin, C. E., Safrit, J. T., Mittler, J., Weinberger, L., Kostrikis, L. G., Zhang, L., Perelson, A. S. and Ho, D. D. (1999). Dramatic rise in plasma viremia after CD8(+) T cell depletion in simian immunodeficiency virus-infected macaques. *J. Exp. Med.*, **189**, 991–8.

Kagi, D., Ledermann, B., Burki, K., Zinkernagel, R. M. and Hengartner, H. (1996). Molecular mechanisms of lymphocyte-mediated cytotoxicity and their role in immunological protection and pathogenesis *in-vivo*. *Ann. Rev. Immunol.*, **14**, 207–32.

Kepler, T. B. and Perelson, A. S. (1993). Cyclic re-entry of germinal center B cells and the efficiency of affinity maturation. *Immunol. Today*, **14**, 412–15.

Kilby, J. M., Hopkins, S., Venetta, T. M., DiMassimo, B., Cloud, G. A., Lee, J. Y., Alldredge, L., Hunter, E., Lambert, D., Bolognesi, D., Matthews, T., Johnson, M. R., Nowak, M. A., Shaw, G. M. and Saag, M. S. (1998). Potent suppression of HIV-1 replication in humans by T-20, a peptide inhibitor of gp41-mediated virus entry. *Nat. Med.*, **4**, 1302–7.

Kimura, M. (1983). *The neutral theory of molecular evolution*, Cambridge University Press.

Kira, J., Koyanagi, Y., Yamada, T., Itoyama, Y., Goto, I., Yamamoto, N., Sasaki, H. and Sakaki, Y. (1991). Increased HTLV-I proviral DNA in HTLV-I-associated myelopathy—a quantitative polymerase chain-reaction study. *Ann. Neurol.*, **29**, 194–201.

Kira, J., Nakamura, M., Sawada, T., Koyanagi, Y., Ohori, N., Itoyama, Y., Yamamoto, N., Sakaki, Y. and Goto, I. (1992). Antibody-titers to HTLV-I-P40TAX protein and GAG-ENV hybrid protein in HTLV-I-associated myelopathy tropical spastic paraparesis—correlation with increased HTLV-I proviral DNA load. *J. Neurol. Sci.*, **107**, 98–104.

Kirschner, D. E., Mehr, R. and Perelson, A. S. (1998). Role of the thymus in pediatric HIV-1 infection. *J. Acquir. Immune Defic. Syndr. Hum. Retrovirol.*, **18**, 95–109.

Klein, J. and Horejsi, V. (1997). *Immunology*, Blackwell Science.

Klenerman, P., Rowland-Jones, S., McAdam, S., *et al.* (1994). Cytotoxic T cell activity antagonized by naturally occurring HIV-1 gag variants. *Nature*, **369**, 403–7.

Klenerman, P., Meier, U. C., Phillips, R. E. and McMichael, A. J. (1995). The effects of natural altered peptide ligands on the whole blood cytotoxic T lymphocyte response to human immunodificiency virus. *Europ. J. Immunol.*, **25**, 1927–31.

Klenerman, P., Phillips, R. E., Rinaldo, C. R., *et al.* (1996a). Cytotoxic T lymphocytes and viral turnover in HIV type 1 infection. *Proc. Natl. Acad. Sci. USA*, **93**, 15323–8.

Klenerman, P., Phillips, R. and McMichael, A. (1996b). Cytotoxic T cell antagonism in HIV. *Sem. Virol.*, **7**, 31–9.

Koenig, S., Gendleman, H. E., Orenstein, J. M., Dal Canto, M. C., Pezeshkpour, G. H., Yungbluth, M., Janotta, F., Aksamit, A., Martin M. A. and Fauci, A. S. (1986). Detection of AIDS virus in microphages in brain tissue from AIDS patients with encephalopathy. *Science*, **233**, 1089–93.

Koenig, S., Earl, P., Powell, D., Pantaleo, G., Merli, S., Moss, B. and Fauci, A. S. (1988). Group-specific major histocompatability complex class-I-restricted cytotoxic responses to human immunodeficiency virus-1 (HIV-1) envelope proteins by cloned peripheral-blood T-cells from an HIV-infected individual. *Proc. Natl. Acad. Sci. USA*, **85**, 8638–42.

Koenig, S., Woods, R. M., Brewah, Y. A., Newell, A. J., Jones, G. M., Boone, E., Adelsberger, J. W., Baseler, M. W., Robinson, S. M. and Jacobson, S. (1993). Characterization of MHC class I restricted cytotoxic T cell responses to tax in HTLV-1 infected patients with neurologic disease. *J. Immunol.*, **151**, 3874–83.

Koot, M., van't Wout, A. B., Koostra, N. A., deGoede, R. E. Y., Tersmette, M. and Schuitemaker, H. (1996). Relation between changes in cellular load, evolution of viral phenotype and the clonal composition of virus populations in the course of human immunodeficiency virus type-1 infection. *J. Infect. Dis.*, **173**, 349–54.

Koup, R. A., Safrit, J. T., Cao, Y. Z., Andrews, C. A., McLeod, G., Borkowski, W., Farthing, C. and Ho, D. D. (1994). Temporal association of cellular immune responses with the initial control of viremia in primary human immunodeficiency virus type-1 syndrome. *J. Virol.*, **68**, 4650–5.

Krakauer, D. C. and Payne, R. J. H. (1997). The evolution of virus induced apoptosis. *Proc. Roy. Soc. Lond B*, **264**, 1757–62.

Krakauer, D. C. and Nowak, M. (1999). T-cell induced pathogenesis in HIV: bystander effects and latent infection. *Proc. R. Soc. Lond. B. Biol. Sci.*, **22**, 1069–75.

Kuiken, C. L., Dejong, J. J., Baan, E., Keulen, W., Tersmette, M. and Goudsmit, J. (1992). Evolution of the V3 envelope domain in proviral sequences and isolates of human immunodeficiency virus type-1 during transition of the viral biological phenotype. *J. Virol.*, **66**, 4622–7.

Larder, B. A. and Kemp, S. D. (1989). Multiple mutations in HIV-1 reverse transcriptase confer high-level resistance to zidovudine (AZT). *Science*, **246**, 1155–8.

Larder, B. A., Darby, G. and Richman, D. D. (1989). HIV with reduced sensitivity to zidovudine (AZT) isolated during prolonged therapy. *Science*, **243**, 1731–4.

Larder, B. A., Kellam, P. and Kemp, S. D. (1993). Convergent combination therapy can select variable multidrug-resistant HIV-1 *in vitro*. *Nature*, **365**, 451–3.

Larder, B. A., Kemp, S. D. and Harrigan, P. R. (1995). Potential mechanism for sustained antiretroviral efficacy of AZT–3TC combination therapy. *Science*, **269**, 696–9.

Laskey, L. A., Groopman, J. E., Fennie, C. W., Benz, P. M., Capon, D. J., Dowbenko, D. J., Nakamura, G. R., Nunes, W. M., Renz, M. E. and Berman, P. W. (1986). Neutralization of the AIDS retrovirus by antibodies to a recombinant envelope glycoprotein. *Science*, **233**, 209–12.

Levin, B. R. and Bull, J. J. (1994). Short-sighted evolution and the virulence of pathogenic microorganisms. *Trends in Microbiology*, **2**, 76–81.

Levin, B. R., Lipsitch, M. and Bonhoeffer, S. (1999). Population biology and infectious disease. *Science*, **283**, 806–9.

Levin, S. A., Grenfell, B., Hastings, A. and Perelson, A. S. (1997). Mathematical and computational challenges in population biology and ecosystem science. *Science*, **275**, 334–43.

Levine, A. (1992). *Viruses*, Scientific American Library, New York.

Levy, J. A. (1998). *HIV and the pathogenesis of AIDS*. 2nd edn. ASM Press.

Levy, J. A., Hoffman, A. D., Kramer, S. M., Landis, J. A., Shimabukuro, J. M. and Oshiro, L. S. (1984). Isolation of lymphocytopathic retroviruses from San Francisco patients with AIDS. *Science*, **225**, 840–2.

Lifson, J. D., Reyes, G. R., McGrath, M. S., Stein, B. S. and Engleman, E. G. (1986*a*). AIDS retrovirus induced cytopathology: giant cell formation and involvement of CD4 antigen. *Science*, **232**, 1123–7.

Lifson, J. D., Feinberg, M. B., Reyes, G. R., Rabin, L., Banapour, B., Chakrabarti, S., Moss, B., Wong-Staal, F. Steimer, K. S. and Engleman, E. G. (1986*b*). Induction of CD4-dependent cell fusion by the HTLV-III/LAV envelope. *Nature*, **323**, 725–8.

Lifson, J. D., Nowak, M. A., Goldstein, S., Rossio, J. L., Kinter, A., Vasquez, G., Wiltrout, T. A., Brown, C., Schneider, D., Wahl, L. M., Lloyd, A. L., Williams, J., Elkins, W. R., Fauci, A. S. and Hirsch, V. M. (1997). The extent of early viral replication is a critical determinant of the natural history of simian immunodeficiency virus (SIV) infection. *J. Virol.*, **71**, 9508–14.

Lifson, J. D., Rossio, J. L., Arnaout, R., Li, L., Parks, T. L., Schneider, D. K., Kiser, R. F., Coalter, V. J., Walsh, G., Imming, R. J., Fisher, B., Flynn, B. M., Bischofberger, N., Piatak, Jr., M., Hirsch, V. M., Nowak, M. A. and Wodarz, D. (2000). Containment of simian immunodeficiency virus infection: cellular immune responses and protection from rechallenge following transient postinoculation antiretroviral treatment. *J. Virol.*, **74**, 2584–93.

Lin, Y. L. and Askonas, B. A. (1981). Biological properties of an influenza A virus-specific killer T cell clone. Inhibition of virus replication *in vivo* and induction of delayed-type hypersensitivity reactions. *J. Exp. Med.*, **154**, 225–34

Little, S. J., McLean, A. R., Spina, C. A., Richman, D. D. and Havlir, D. V. (1999). Viral dynamics of acute HIV-1 infection. *J. Exp. Med.*, **190**, 841–50.

London, W. T. and Blumberg, B. S. (1982). A cellular model of the role of hepatitis B virus in the pathogenesis of primary hepatocellular carcinoma. *Hepatology*, **2**, S10–14.

Loveday, C., Kay, S., Tenant-Flowers, S., *et al.* (1995). HIV-1 RNA serum load and resistant viral genotypes during early zidovudine therapy. *The Lancet*, **345**, 820–4.

Lukashov, V. V., Kuiken, C. L. and Goudsmit, J. (1995). Intrahost human immunodeficiency virus type 1 evolution is related to length of the immunocompetent period. *J. Virol.*, **69**, 6911–16.

Ljunggren, K., Biberfeld, G., Jondal, M. and Fenyo, E. M. (1989). Antibody-dependent cellular cytotoxicity detects type- and strain-specific antigens among human immunodeficiency virus types 1 and 2 and simian immunodeficiency virus SIVmac isolates. *J. Virol.*, **63**, 3376–81.

Mansky, L. M. and Temin, H. M. (1995). Lower *in vivo* mutation rate of human immunodeficiency virus type 1 than the predicted from the fidelity of purified reverse transcriptase. *J. Virol.*, **29**, 5087–94.

Marchuk, G. I. (1997). Mathematical modelling of immune response in infectious diseases. In *Mathematics and its applications*, Vol. 395. Kluwer Academic Publishers.

Marchuk, G. I., Petrov, R. V., Romanyukha, A. A. and Bocharov, G. A. (1991*a*). Mathematical model of antiviral immune response. I. Data analysis, generalized picture construction and parameters evaluation for hepatitis B. *J. Theor. Biol.*, **151**, 1–40.

Marchuk, G. I., Romanyukha, A. A. and Bocharov, G. A. (1991*b*). Mathematical model of antiviral immune response. II. Parameters identification for acute viral hepatitis B. *J. Theor. Biol.*, **151**, 41–69.

Markowitz, M., Mo, H., Kempf, D. J., Norbeck, D. W., Bhat, T. N., Erickson, J. W. and Ho, D. D. (1995). Selection and analysis of human immunodeficiency virus type 1 variants with increased resistance to ABT-538, a novel protease inhibitor. *J. Virol.*, **69**, 701–6.

Martin, M. P., Dean, M., Smith, M. W., Winkler, C., Gerrard, B., Michael, N. L., Lee, B., Doms, R. W., Margolick, J., Buchbinder, S., Goedert, J. J., O'Brien, T. R., Hilgartner, M. W., Vlahov, D., O'Brien, S. J. and Carrington, M. (1998). Genetic acceleration of AIDS progression by a promoter variant of CCR5. *Science*, **282**, 1907–11.

Matloubian, M., Concepcion, R. J. and Ahmed, R. (1994). CD4+ cells are required to sustain CD8+ cytotoxic T-cell responses during chronic viral infection. *J. Virol.*, **68**, 8056–63.

May, R. M. (1973). *Stability and complexity in model ecosystems.* Princeton University Press.

May, R. M. and Anderson, R. M. (1979). Population biology of infectious diseases. II. *Nature*, **280**, 455.

May, R. M. and Nowak, M. A. (1994). Superinfection, metapopulation dynamics, and the evolution of diversity. *J. Theor. Biol.*, **170**, 95–114.

May, R. M. and Nowak, M. A. (1995). Coinfection and the evolution of parasite virulence. *Proc. Roy. Soc. Lond. B.*, **261**, 209–215.

Maynard Smith, J. (1970). Natural selection and the concept of protein space. *Nature*, **225**, 563–4.

Maynard Smith, J. (1982). *Evolution and the theory of games*, Cambridge University Press.

Maynard Smith, J. (1998). *Evolutionary genetics*, Oxford University Press.

McAdam, S. N., Klenerman, P., Tussey, L. G., Rowland-Jones, S., Lalloo, D., Brown, A. L., *et al.* (1995). Immunogenic HIV variant peptides that bind to HLA-B8 but fail to stimulate cytotoxic T lymphocyte responses. *J. Immunol.*, **155**, 2729–36.

McCaskill, J. S. (1984*a*). A stochastic theory of macromolecular evolution. *Biol. Cybernet.*, **50**, 63–73.

McCaskill, J. S. (1984*b*). A localization threshold for macromolecular quasispecies from continuously distributed replication rates. *J. Chem. Phys.*, **80**, 5194–202.

McChesney, M. B. and Oldstone, M. B. A. (1987). Viruses perturb lymphocyte functions: selected principles characterizing virus-induced immunosuppression. *Ann. Rev. Immunol.*, **5**, 279–304.

McElrath, M. J., Rabin, M., Hoffman, M., Klucking, S., Garcia, J. V. and Greenberg, P. D. (1994). Evaluation of human-immunodeficiency-virus type-1 (HIV-1)-specific cytotoxic T-lymphocyte responses utilizing B-lymphoblastoid cell-lines transduced with the CD4 gene and infected with HIV-1. *J. Virol..* **68**, 5074–83.

McKnight, A. and Clapham, P. R. (1995). Immune escape and tropism for HIV. *TIM*, **3**, 356–61.

McLean, A. R. (1993). The balance of power between HIV and the immune system. *Trend. Microbiol.*, **1**, 9–13.

McLean, A. R. and Frost, S. D. W. (1995). Zidovudine and HIV: mathematical models of within-host population dynamics. *Rev. Med. Virol.*, **5**, 141–7.

McLean, A. R. and Kirkwood, T. B. L. (1990). A model of human immunodeficiency virus infection in T helper cell clones. *J. Theor. Biol.*, **147**, 177–203.

McLean, A. R. and Mitchie, C. A. (1995). *In vivo* estimates of division and death rates of human T lymphocytes. *Proc. Natl. Acad. Sci. USA*, **92**, 3707–11.

McLean, A. R. and Nowak, M. A. (1992*a*). Competition between zidovudine-sensitive and zidovudine-resistant strains of HIV. *AIDS*, **6**, 71–9.

McLean, A. R. and Nowak, M. A. (1992*b*). Models of interactions of HIV and other pathogens. *J. Theor. Biol.*, **155**, 69–102.

McLean, A. R., Emery, V. C., Webster, A. and Griffiths, P. D. (1991). Population dynamics of HIV within an individual after treatment with zidovudine. *AIDS*, **5**, 485–9.

McLean, A. R., Rosado, N. M., Agenes, F., Vasconcellos, R. and Freitas, A. A. (1997). Resource competition as a mechanism for B cell homeostasis. *Proc. Natl. Acad. Sci. USA*, **94**, 5792–7.

McMichael, A. J. (1993). Natural selection at work on the surface of virus-infected cells. *Science*, **260**, 1771–2.

McMichael, A. J. and Phillips, R. E. (1997). Escape of human immunodeficiency virus from immune control. *Ann. Rev. Immunol.*, **15**, 271–96.

McMichael, A., Rowland-Jones, S., Klenerman, P., McAdam, S., Gotch, F., Phillips, R. and Nowak, M. (1995). Epitope variation and T-cell recognition. *J. Cell. Biochem.*, **60**, 60.

McMichael, A., Goulder, P., Rowlandjones, S., Nowak, M. and Phillips, R. (1996). HIV escapes from cytotoxic lymphocytes. *Immunology*, **89**, SE111.

McNearney, T., Hornickova, Z., Markham, R., Birdwell, A., Arens, M., Saah, A. and Ratner, L. (1992). Relationship of human immunodeficiency virus type-1 sequence heterogeneity to stage of disease. *Proc. Natl. Acad. Sci. USA*, **89**, 10247–51.

Mehr, R., Perelson, A. S., Fridkis-Hareli, M. and Globerson, A. (1996). Feedback regulation of T cell development in the thymus. *J. Theor. Biol.*, **181**, 157–67.

Mellors, J. W., Kingsley, L. A., Rinaldo, Jr., C. R., Todd, J. A., Hoo, B. S., Kokka, R. P. and Gupta, P. (1995). Quantitation of HIV-1 RNA in plasma predicts outcome after seroconversion. *Ann. Intern. Med.*, **122**, 573–9.

Mellors, J. W., Rinaldo, Jr., C. R., Gupta, P., White, R. M., Todd, J. A. and Kingsley, L. A. (1996). Prognosis in HIV-1 infection predicted by the quantity of virus in plasma. *Science*, **272**, 1167–70.

Meyerhans, A., Cheynier, R., Albert, J., Seth, M., Kwok, S., Sninsky, J., Morfeldt-Manson, L., Asjo, B. and Wain-Hobson, S. (1989). Temporal fluctuations in HIV quasispecies *in vivo* are not reflected by sequential HIV isolations. *Cell*, **58**, 901–10.

Mills, D. R., Kramer, F. R. and Spiegelman, S. (1973). Complete nucleotide sequence of a replicating RNA molecule. *Science*, **180**, 916–18.

Mittler, J. E., Levin, B. R., Antia, R. (1996). T-cell homeostosis, competition and drift. *J. Ac. Im. Def. Syn.*, **12**, 233–48.

Mittler, J. E., Sulzer, B., Neumann, A. U. and Perelson, A. S. (1998). Influence of delayed viral production on viral dynamics in HIV-1 infected patients. *Math. Biosci.*, **152**, 143–63.

Mohri, H., Singh, M. K., Ching, W. T. and Ho D. D. (1993). Quantitation of zidovudine-resistant human immunodeficiency virus type 1 in the blood of treated and untreated patients. *Proc. Natl. Acad. Sci. USA*, **90**, 25–9.

Mohri, H., Bonhoeffer, S., Monard, S., Perelson, A. S. and Ho, D. D. (1998). Rapid turnover of T lymphocytes in SIV-infected rhesus macaques. *Science*, **279**, 1223–7.

Moore, J. P. and Nara, P. L. (1991). The role of the V3 loop of gp120 in HIV infection. *AIDS*, **5**(Suppl. 2), S21–33.

Moskophidis, D. and Zinkernagel, R. M. (1995). Immunobiology of cytotoxic T-cell escape mutants of lymphocytic choriomeningitis virus. *J. Virol.*, **69**, 2187–93.

Mullins, J. I., Hoover, E. A. and Quackenbush, S. L. (1991). Disease progression and viral genome variants in experimental feline leukemia virus-induced immunodeficiency syndrome. *J. Acq. Immun. Defic. Synd. Hum. R*, **4**, 547–57.

Munoz, A., Vlahov, D., Solomon, L., Margolick, J. B., Bareta, J. C., Cohn, S., Astemborski, J. and Nelson, K. E. (1992). Prognostic indicators for development of AIDS among intravenous drug users. *J. Acquir. Immune Defic. Syndr.*, **5**, 694–700.

Murray, J. D. (1993). *Mathematical biology*. 2nd corr. edn. Springer-Verlag.

Nakamoto, Y., Guidotti, L. G., Kuhlen, C. V., Fowler, P. and Chisari, F. V. (1998). Immune pathogenesis of hepatocellular carcinoma. *J. Exp. Med.*, **188**, 341–50.

Nagai, M., Usuku, K., Matsumoto, W., Kodama, D., Takenouchi, N., Moritoyo, T., Hashigichi, S. and Ichinose, M. (1998). Analysis of HTLV-I proviral load in 202 HAM/TSP patients and 243 asymptomatic HTLV-I carriers: high proviral load strongly predisposes to HAM/TSP. *J. Neurovirol.*, **4**, 586–93.

Nájera, I., Holguin, A. and Quinones-Mateu, M. E., *et al.* (1995). *pol* gene quasispecies of human immunodeficiency virus: mutations associated with drug resistance in virus from patients undergoing no drug therapy. *J. Virol.*, **69**, 23–31.

Nara, P. L., Smit, L., Dunlop, N., Hatch, W., Merges, M., Waters, D., Kelliher, J., Gallo, R. C., Fischinger, P. J. and Goudsmit, J. (1990). Emergence of virus resistant to neutralization by V3-specific antibodies in experimental HIV-1-IIIB infection of chimpanzees. *J. Virol.*, **64**, 3779–91.

Nelson, G. W. and Perelson, A. S. (1992). A mechanism of immune escape by slow-replicating HIV strains. *J. Acquir. Immune Defic. Syndr.*, **5**, 82–93.

Neumann, A. U., Lam, N. P., Dahari, H., Gretch, D. R., Wiley, T. E., Layden T. J. and Perelson, A. S. (1998). Hepatitis C viral dynamics in *vivo* and the antiviral efficacy of interferon-alpha therapy. *Science*, **282**, 103–7.

Niewiesk, S. and Bangham, C. R. M. (1996). Evolution in a chronic RNA virus infection: selection on HTLV-I tax protein differs between healthy carriers and patients with tropical spastic paraparesis. *J. Molec. Evol.*, **42**, 452–8.

Niewiesk, S., Daenke, S., Parker, C. E., Taylor, G., Weber, J., Nightingale, S. and Bangham, C. R. (1994). The transactivator gene of human T-cell leukemia virus type I is more variable within and between healthy carriers than patients with tropical spastic paraparesis. *J. Virol.*, **68**, 6778–81.

Niewiesk, S., Daenke, S., Parker, C. E., Taylor, G., Weber, J., Nightingale, S. and Bangham, C. R.(1995). Naturally occurring variants of human T-cell leukemia virus type I Tax protein impair its recognition by cytotoxic T lymphocytes and the transactivation function of Tax. *J. Virol.*, **69**, 2649–53.

Nowak, M. (1990). HIV mutation-rate. *Nature*, **347**, 522.

Nowak, M. A. (1992*a*). Variability of HIV infections. *J. Theor. Biol.*, **155**, 1–20.

Nowak, M. A. (1996). Immune responses against multiple epitopes: a theory for immunodominance and antigenic variation. *Semi. Virol.*, **7**, 83–92.

Nowak, M. A. and Bangham, C. R. M. (1996). Population dynamics of immune responses to persistent viruses. *Science*, **272**, 74–9.

Nowak, M. A. and May, R. M. (1991). Mathematical biology of HIV infections—antigenic variation and diversity threshold. *Math. Biosci.*, **106**, 1–21.

Nowak, M. A. and May, R. M. (1992). Coexistence and competition in HIV infections. *J. Theor. Biol.*, **159**, 329–42.

Nowak, M. and May, R. M. (1993). AIDS Pathogenesis—mathematical models of HIV and SIV infections. *AIDS*, **7**, S3–18.

Nowak, M. A. and May, R. M. (1994). Superinfection and the evolution of virulence. *Proc. Roy. Soc. Lond. B.*, **255**, 81–89 (installer).

Nowak, M. A. and McLean, A. R. (1991). A mathematical model of vaccination against HIV to prevent the development of AIDS. *Proc. Roy. Soc. Lond. B.*, **246**, 141–6.

Nowak, M. A. and McMichael, A. J. (1995). How HIV defeats the immune system. *Sci. Am., August*, 42–9

Nowak, M. and Schuster, P. (1989). Error thresholds of replication in finite populations mutation frequencies and the onset of Muller's ratchet. *J. Theor. Biol.*, **137**, 375–95.

Nowak, M. A., May, R. M. and Anderson R. M. (1990). The evolutionary dynamics of HIV-1 quasispecies and the development of immunodeficiency disease. *AIDS*, **4**, 1095–103.

Nowak, M. A., Anderson, R. M., McLean, A. R., Wolfs, T. F. W., Goudsmit, J. and May, R. M. (1991). Antigenic diversity thresholds and the development of AIDS. *Science*, **254**, 963–9.

Nowak, M. A., May, R. M., Phillips, R. E., *et al.* (1995*a*). Antigenic oscillations and shifting immunodominance in HIV-1 infections. *Nature*, **375**, 606–11.

Nowak, M. A., May, R. M. and Sigmund, K. (1995*b*). Immune-responses against multiple epitopes. *J. Theor. Biol.*, **175**, 325–53.

Nowak, M. A., Bonhoeffer, S., Loveday, C., *et al.* (1995*c*). HIV dynamics: results confirmed. *Nature*, **375**, 193.

Nowak M. A., Anderson, R. M., Boerlijst, M. C., Bonhoeffer S., May, R. M. and McMichael, A. J. (1996*a*). HIV-1 evolution and disease progression. *Science*, **274**, 1008–11.

Nowak, M. A., Bonhoeffer, S., Hill, A. M., Boehme, R., Thomas, H. C. and McDade, H. (1996*b*). Viral dynamics in hepatitis B virus infection. *Proc. Natl. Acad. Sci. USA*, **93**, 4398–402.

Nowak, M. A., Bonhoeffer, S., Shaw, G. M., *et al.* (1997*a*). Anti-viral drug treatment: dynamics of resistance in free virus and infected cell populations. *J. Theor. Biol.*, **184**, 203–17.

Nowak, M. A., Lloyd, A. L., Vasquez, G., Wiltrout, T. A., Wahl, L. M., Bischofsberger, N., Williams, J., Kinter, A., Fauci, A. S., Hirsch, V. M. and Lifson, J. D. (1997*b*). Viral dynamics of primary viremia and antiviral therapy in simian immunodeficiency virus infection. *J. Virol.*, **71**, 7518–25.

O'Brien, S. J. and Dean, M. (1997). In search of AIDS resistance genes. *Sci. Am., September*, **277**, 28–35.

O'Brien, W. A. (1994). HIV-1 entry and reverse transcription in macrophages. *J. Leukocyte Biol.*, **56**, 273–7.

Ogg, G. S., Jin, X., Bonhoeffer, S., Dunbar, P. R., Nowak, M. A., Monard, S., Segal, J. P., Cao, Y., Rowland-Jones, S. L., Cerundolo, V., Hurley, A., Markowitz, M., Ho, D. D., Nixon, D.F. and McMichael, A. J. (1998). Quantitation of HIV-1-specific cytotoxic T lymphocytes and plasma load of viral RNA. *Science*, **279**, 2103–6.

Ogg, G. S., Jin, X., Bonhoeffer, S., Moss, P., Nowak, M. A., Monard, S., Segal, J. P., Cao, Y., Rowland-Jones, S. L., Hurley, A., Markowitz, M., Ho, D. D., McMichael, A. J. and Nixon, D. F. (1999). Decay kinetics of human immunodeficiency virus-specific effector cytotoxic T lymphocytes after combination antiretroviral therapy. *J. Virol.*, **73**, 797–800.

Ostrowski, M. A., Krakauer, D. C., Li, Y., Justement, S. J., Learn, G., Ehler, L. A., Stanley, S. K., Nowak, M. and Fauci, A. S. (1998). Effect of immune activation on the dynamics of human immunodeficiency virus replication and on the distribution of viral quasispecies. *J. Virol.*, **72**, 7772–84.

Overbaugh, J., Rudensey, L. M., Papenhausen, M. D., Benveniste, R. E. and Morton, W. R. (1991). Variation in SIV *env* is confined to V1 and V4 during progression to simian AIDS. *J. Virol.*, **65**, 7025–31.

Parker, C. E., Nightingale, S., Taylor, G. P., Weber, J. and Bangham, C. R. M. (1994). Circulating anti-Tax cytotoxic T lymphocytes from human T-cell leukemia virus type I-infected people, with and without tropical spastic paraparesis, recognize multiple epitopes simultaneously. *J. Virol.*, **68**, 2860–8.

Parvin, J. D., Moscona, A., Pan W. T., *et al.* (1986). Measurement of the mutation rates of animal viruses influenza A virus and poliovirus type 1. *J. Virol.*, **59**, 377–83.

Payne, R. J. H., Nowak, M. A. and Blumberg, B. S. (1996). The dynamics of hepatitis B virus infection. *Proc. Natl. Acad. Sci. USA*, **93**, 6542–6.

Pelletier, E., Saurin, W., Cheynier, R., Letvin, N. L. and Wain-Hobson, S. (1995). The tempo and mode of SIV quasispecies development *in vivo* calls for massive viral replication and clearance. *Virology*, **208**, 644–52.

Penna, A., Artini, M., Cavalli, A., Levrero, M., Bertoletti, A., Pilli, M., Chisari, F. V., Rehermann, B., Del Prete, G., Fiaccadori F. and Ferrari, C. (1996). Long-lasting memory T cell responses following self-limited acute hepatitis B. *J. Clin. Invest.*, **98**, 1185–94.

Perelson, A. S. and Bell, G. I. (1982). Delivery of lethal hits by cytotoxic T lymphocytes in multicellular conjugates occurs sequentially but at random times. *J. Immunol.*, **129**, 2796–801.

Perelson, A. S. and Oster, G. F. (1979). Theoretical studies of clonal selection: minimal antibody repertoire size and reliability of self-non-self discrimination. *J. Theor. Biol.*, **81**, 645–70.

Perelson, A. S. and Weisbuch, G. (1992). Modeling immune reactivity in secondary lymphoid organs. *Bull. Math. Biol.*, **54**, 649–72.

Perelson, A. S., Mirmirani, M. and Oster, G. F. (1976). Optimal strategies in immunology I. B-cell differentation and proliferation. *J. Math. Biol.*, **3**, 325–67.

Perelson, A. S., Mirmirani, M. and Oster, G. F. (1978). Optimal strategies in immunology II. B memory cell production. *J. Math. Biol.*, **5**, 213–56.

Perelson, A. S., Goldstein, B. and Rocklin, S. (1980). Optimal strategies in immunology III. The IgM–IgG switch. *J. Math. Biol.*, **10**, 209–56.

Perelson, A. S., Kirschner, D. E. and de Boer, R. (1993). Dynamics of HIV infection of CD4+ T cells. *Math. Biosci.*, **114**, 81–125.

Perelson, A. S., Neumann, A. U., Markowitz, M., Leonard, J. M. and Ho, D. D. (1996). HIV-1 dynamics *in-vivo*—virion clearance rate, infected cell life-span and viral generation time. *Science*, **271**, 1582–6.

Perelson, A. S., Essunger, P., Cao, Y., *et al.* (1997*a*). Decay characteristsics of HIV-1 infected compartments during combination therapy. *Nature*, **387**, 188–91.

Perelson, A. S., Essunger, P. and Ho, D. D. (1997*b*). Dynamics of HIV-1 and CD4+ lymphocytes *in vivo*. *AIDS*, **11**, S17–24.

Phillips A. N. (1996). Reduction of HIV concentration during acute infection: independence from a specific immune response. *Science*, **271**, 497–9.

Phillips, A. N., McLean, A. R., Loveday, C., Tyrer, M., Bofill, M., Devereux, H., Madge, S., Dykoff, A., Drinkwater, A., Burke, A., Huckett, L., Janossy, G. and Johnson, M. A. (1999). *In vivo* HIV-1 replicative capacity in early and advanced infection. *AIDS*, **13**, 67–73.

Phillips, R. E., Rowland-Jones, S., Nixon, D. F., *et al.* (1991). Human immunodeficiency virus genetic variation that can escape cytotoxic T cell recognition. *Nature*, **354**, 453–9.

Piatak, Jr., M., Saag, M. S., Yang, L. C., Clark, S. J., Kappes, J. C., Luk, K. C., Hahn, B. H., Shaw, G. M. and Lifson, J. D. (1993). High levels of HIV-1 in plasma during all stages of infection determined by competitive PCR. *Science*, **259**, 1749–54.

Pircher, H., Moskophidis, D., Rohrer, U., Buerki, K., Hengartner, H. and Zinkernagel, R. M. (1990). Viral escape by selection of cytotoxic T cell-resistant virus variants *in vivo*. *Nature*, **346**, 629–33.

Plata, F., Autran, B., Martins, L. P., Wain-Hobson, S., Raphael, M., Mayaud, C., Denis, M., Guillon, J. M. and Debre, P. (1987). AIDS virus-specific cytotoxic lymphocytes-T in lung disorders. *Nature*, **328**, 348–51.

Preston, B. D., Poiesz, B. J. and Loeb, L. A. (1988). Fidelity of HIV-1 reverse-transcriptase. *Science*, **242**, 1168–71.

Price, D. A., Goulder, P. J. R., Klenerman, P., Sewell, A. K., Easterbrook, P. J., Troop, M., *et al.* (1997). Positive selection of HIV-1 cytotoxic T lymphocyte escape variants during primary infection. *Proc. Natl. Acad. Sci. USA*, **94**, 1890–5.

Ramratnam, B., Mittler, J. E., Zhang, L., Boden, D., Hurley, A., Fang, F., Macken, C. A., Perelson, A. S., Markowitz, M. and Ho, D. D. (2000). The decay of the latent reservoir of replication-competent HIV-1 is inversely correlated with the extent of residual viral replication during prolonged anti-retroviral therapy. *Nat. Med.*, **6**, 82–5.

Regoes, R., Wodarz, D. and Nowak, M. A. (1998). The effect of target cell limitation and immune responses on virus evolution. *J. Theor. Biol.*, **191**, 451–62.

Rehermann, B., Ferrari, C., Pasquinelli, C. and Chisari, F. V. (1996). The hepatitis-B virus persists for decades after patients recovery from acute viral hepatitis despite active maintenance of a cytotoxic T-lymphocyte response. *Nat. Med.*, **2**, 1104–8.

Ribeiro, R. M., Bonhoeffer, S. and Nowak, M. A. (1998). The frequency of resistant mutant virus before anti-viral therapy. *AIDS*, **12**, 461–5.

Richman, D. D. (1990). Zidovudine resistance of human immunodeficiency virus. *Rev. Infect. Dis.*, **12**(Suppl. 5), S507–10.

Richman, D. D. (1994*a*). Drug resistance in viruses. *Trend. Microbiol.*, **2**, 401–7.

Richman, D. D. (1994*b*). Resistance, drug failure and disease progression *AIDS Res. Hum. Retrovir.*, **10**, 901–5.

Richman, D. D., Havlir, D., Corbeil, J., Looney, D., Ignacio, C., Spector, S. A., *et al.* (1994). Nevirapine resistance mutations of human immunodeficiency virus type 1 selected during therapy. *J. Virol.*, **68**, 1660–6.

Rinaldo, C. R., Beltz, L. A., Huang, X. L., Gupta, P., Fan, Z. and Torpey, D. J. (1995*a*). Anti-HIV type I cytotoxic T lymphocyte effector activity and disease progression in the first 8 years of HIV type 1 infection of homosexual men. *AIDS Res. Hum. Retrovir.*, **11**, 481–9.

Rinaldo, C. R., Huang, X. L., Fan, Z., *et al.* (1995*b*). High levels of anti-human immunodeficiency virus type 1 (HIV-1) memory cytotoxic T-lymphocyte activity and low viral load are associated with lack of disease in HIV-1 infected long-term nonprogressors. *J. Virol.*, **69**, 5838–42.

Robert-Guroff, M., Brown, M. and Gallo, R. C. (1985). HTLV-III-neutralizing antibodies in patients with AIDS and AIDS-related complex. *Nature*, **316**, 72–4.

Robert-Guroff, M., Reitz, Jr., M. S., Robey, W. G. and Gallo, R. C. (1986). *In vitro* generation of an HTLV-III variant by neutralizing antibody. *J. Immunol.*, **137**, 3306–9.

Roberts, J. D., Bebenek, K. and Kunkel, T. A. (1988). The accuracy of reverse trascriptase from HIV-1. *Science*, **242**, 1171–3.

Roitt, I. M., Brostoff, J. and Male, D. K. (1998). *Immunology*, Mosby.

Rowland-Jones, S. and Tan, R. (1997). Control of HIV coreceptor expression: implications for pathogenesis and treatment. *Trend. Microbiol.*, **5**, 300–3.

Saag, M. S., Hahn, B. H., Gibbons, J., Li, Y. X., Parks, E. S., Parks, W. P. and Shaw, G. M. (1988). Extensive variation of human immunodeficiency virus type-1 *in vivo*. *Nature*, **334**, 440–4.

St. Clair, M. H., Martin, J. L., Tudor-Williams, G., Bach, M. C., Vavro, C. L., King, D. M., *et al.* (1991). Resistance to ddI and sensitivity to AZT induced by a mutation in HIV-1 reverse transcriptase. *Science*, **253**, 1557–9.

Salinovich, O., Payne, S. L., Montelaro, R. C., Hussain, K. A., Issel, C. J. and Schnorr, K. L. (1986). Rapid emergence of novel antigenic and genetic variants of EIAV during persistant infection. *J. Virol.*, **57**, 71–80.

Samson, M., Libert, F., Doranz, B. J., *et al.* (1996). Resistance to HIV-1 infection in caucasian individuals bearing mutant alleles of the CCR-5 chemokine receptor gene. *Nature*, **382**, 722–5.

Sasaki, A. (1994). Evolution of antigen drift/switching: continuously evading pathogens. *J. Theor. Biol.*, **168**, 291–308.

Sattentau, Q. J., Dalgleish, A. G., Weiss, R. A. and Beverley, P. C. (1986). Epitopes of the CD4 antigen and HIV infection. *Science*, **234**, 1120–3.

Schulz, T. F., Whitby, D., Hoad, J. G., Corrah, T., Whittle, H. and Weiss, R. A. (1990). Biological and molecular variability of HIV-2 isolates from The Gambia. *J. Virol.*, **64**, 5177–82.

Schuurman, R., Nijhuis, M., van Leeuwen, R., *et al.* (1995). Rapid changes in human immunodeficiency type 1 RNA load and appearance of drug resistant virus populations in persons treated with lamivudine (3TC). *J. Infect Dis.*, **171**, 1411–19.

Segel, L. A. (1998). Multiple attractors in immunology: theory and experiment. *Biophys. Chem.*, **72**, 223–30.

Segel, L. A. and Bar-Or, R. L. (1999). On the role of feedback in promoting conflicting goals of the adaptive immune system. *J. Immunol.*, **163**, 1342–9.

Segel, L. A. and Jager, E. (1994). Reverse engineering: a model for T-cell vaccination. *Bull. Math. Biol.*, **56**, 687–721.

Segel, L. A. and Perelson, A. S. (1989). Shape space: an approach to the evaluation of cross-reactivity effects, stability and controllabilty in the immune system. *Immunol. Lett.*, **22**, 91–9.

Seiden, P. E. and Celada, F. (1992). A model for simulating cognate recognition and response in the immune system. *J. Theor. Biol.*, **158**, 329–57.

Semper, M. and Luce, R. (1975). Evidence of *de novo* production of self-replicating and environmentally adapted RNA structures by bacteriophage Q-beta replicase. *Proc. Natl. Acad. Sci. USA*, **72**, 162–6.

Sercarz, E. E., Lehmann, P. V., Ametani, A., Benichou, G., Miller, A. and Moudgil, K. (1993). Dominance and crypticity of T-cell antigenic determinants. *Ann. Rev. Immunol.*, **11**, 729–66.

Shankarappa, R., Margolick, J. B., Gange, S. J., Rodrigo, A. G., Upchurch, D., Farzadegan, H., Gupta, P., Rinaldo, C. R., Learn, G. H., He, X., Huang, X. L. and Mullins, J. I. (1999). Consistent viral evolutionary changes associated with the progression of human immunodeficiency virus type 1 infection. *J. Virol.*, **73**, 10489–502.

Shaw, G. M., Hahn, B. H., Arya, S. K., Groopman, J. E., Gallo, R. C. and Wong-Staal, F. (1984). Molecular characterization of human T-cell leukemia (lymphotropic) virus type III in the acquired immune deficiency syndrome. *Science*, **226**, 1165–71.

Shaw, G. M., Harper, M. E., Hahn, B. H., Epstein, L. G., Gajdusek, D. C., Price, R. W., Navia, B. A., Petito, C. K., O'Hara, C. J., Groopman, J. E., *et al.* (1985). HTLV-III infection in brains of children and adults with AIDS encephalopathy. *Science*, **227**, 177–82.

Simmonds, P., Balfe, P., Peutherer, J. F., Ludlam, C. A., Bishop, J. O. and Brown, A. J. L. (1990). Analysis of sequence diversity in hypervariable regions of the external glycoprotein of human immunodeficiency virus type 1. *J. Virol.*, **64**, 5840–50.

Simmonds, P., Zhang, L. Q., McOmish, F., Balfe, P., Ludlam, C. A. and Brown, A. J. L. (1991). Discontinuous sequence change of human immunodeficiency virus (HIV) type 1 env sequences in plasma viral and lymphocyte-associated proviral populations *in vivo*: Implications for models of HIV pathogenesis. *J. Virol.*, **65**, 6266–76.

Smith, M. W., Dean, M., Carrington, M., *et al.* (1997). Contrasting genetic influence of CCR2 and CCR5 variants on HIV-1 infection and disease progression. *Science*, **277**, 959–65.

Spindler, K. R., Horodyski, F. M. and Holland, J. J. (1982). High multiplicities of infection favor rapid and random evolution of vesicular stomatitis virus. *Virology*, **119**, 96–108.

Stafford, *et al.* (2000). *J. Theor. Bio.*, in press.

Staszewski, S. (1995). Zidovudine and lamivudine: results of phase III studies. *J. Aquir. Immune Defic. Syndr. Hum. Retrovirol.*, **10**(Suppl. 1), S57.

Stevenson, M. (1996). Portals of entry—uncovering HIV nuclear transport pathways. *Trend. Cell Biol.*, **6**, 9–15.

Stilianakis, N. I., Boucher, C. A. B., De Jong, M. D., van Leeuwen, R., Schuurman, R. and De Boer, R. J. (1997). Clinical data sets of human immunodeficiency virus type-1 reverse transcriptase resistant mutants explained by a mathematical model. *J. Virol.*, **71**, 161–8.

Swetina, J. and Schuster, P. (1982). Self-replication with errors. A model for polynucleotide replication. *Biophys. Chem.*, **16**, 329–45.

Tersmette, M., deGoede, R. E. Y., Al, B. J. M., Winkel, I. N., Gruters, R. A., Cuypers, H. T., *et al.* (1988). Differential syncytium-inducing capacity of human immunodeficiency virus isolates—frequent detection of syncytium-inducing isolates in patients with acquired immunodeficiency syndrome (AIDS) and AIDS related complex. *J. Virol.*, **62**, 2026–32.

Wahlberg, J. Albert, J., Lundeberg, J., von Gegerfelt, A., Boliden, K., Utter, G., Fenyö, E. -M. and Uhlen, M. (1991). Analysis of the V3 loop in neutralization-resistant human immunodeficiency virus type 1 variants by direct solid-phase DNA sequencing. *AIDS Res. Hum. Retrovir.*, **7**, 983–90.

Wain-Hobson, S. (1989). HIV genome variability *in vivo*. *AIDS*, **3**(Suppl. 1), S13–18.

Wain-Hobson, S. (1993). The fastest genome evolution ever described: HIV variation *in situ*. *Curr. Opin. Genet. Dev.*, **3**, 878–83.

Wain-Hobson, S., Sonigo, P., Danos, O., Cole, S. and Alizon, M. (1985). Nucleotide sequence of the AIDS virus, LAV. *Cell*, **40**, 9–17.

Walker, B. D., Chakrabarti, S., Moss, B., Paradis, T. J., Flynn, T., Durno, A. G., Blumberg, R. S., Kaplan, J. C., Hirsch, M. S. and Schooley, R. T. (1987). HIV-specific cytotoxic lymphocytes-T in seropositive individuals. *Nature*, **328**, 345–8.

Walker, B. D., Flexner, C., Paradis, T. J., Fuller, T. C., Hirsch, M. S., Schooley, R. T. and Moss, B. (1988). HIV-1 reverse-transcriptase is a target for cytotoxic lymphocyte-T in infected individuals. *Science*, **240**, 64–6.

Walker, C. M., Moody, D. J., Stites, D. P. and Levy J. A. (1986). CD8+ lymphocytes can control HIV infection *in vitro* by suppressing virus replication. *Science*, **234**, 1563–6.

Wei, X. P., Ghosh, S. K., Taylor, M. E., *et al.* (1995). Viral dynamics in HIV-1 infection. *Nature*, **373**, 117–22.

Wein, L. M., Zenios, S. and Nowak, M. A. (1997). Dynamic multidrug therapies for HIV: a control theoretic approach. *J. Theor. Biol.*, **185**, 15–29.

Weiss, R. A., Clapham, P. R., Cheingsongpopov, R., *et al.* (1985a). Neutralization of human T-lymphotropic virus type III by sera of AIDS and AIDS-risk patients. *Nature*, **316**, 69–72.

Weiss, R. A., Clapham, P. R., Weber, J. N., Dalgleish, A. G., Lasky, L. A. and Berman, P. W. (1985*b*). Neutralization of human T-lymphotropic virus type III by sera of AIDS and AIDS-risk patients. *Nature*, **316**, 69–72.

Weiss, R. A., Clapham, P. R., Weber, J. N., Dalgleish, A. G., Lasky, L. A. and Berman, P. W. (1986). Variable and conserved neutralization antigens of human immunodeficiency virus. *Nature*, **324**, 572–5.

Winkler, C., Modi, W., Smith, M. W., Nelson, G. W., Wu, X., Carrington, M., Dean, M., Honjo, T., Tashiro, K., Yabe, D., Buchbinder, S., Vittinghoff, E., Goedert, J. J., O'Brien, T. R., Jacobson, L. P., Detels, R., Donfield, S., Willoughby, A., Gomperts, E., Vlahov, D., Phair, J. and O'Brien, S. J. (1998). Genetic restriction of AIDS pathogenesis by an SDF-1 chemokine gene variant. *Science*, **279**, 389–93.

Wodarz, D. and Nowak, M. A. (1998*a*). Virus dynamics and cell tropism: competition and the evolution of specialism. *Math. Biosci.*, submitted.

Wodarz, D. and Nowak, M. A. (1998*b*). Viral dynamics under different immune responses. *J. Theor. Med.*, submitted.

Wodarz, D. and Nowak, M. A. (1998*c*). The effect of different immune responses on the evolution of virulent CXCR4 tropic HIV. *Proc. Roy. Soc. Lond, B.*, **265**, 2149–58.

Wodarz, D. and Nowak, M. A. (1998*d*). Mathematical models of virus dynamics and resistance. *J. HIV Ther.*, **3**, 36–41.

Wodarz, D. and Nowak, M. A. (1999). Specific therapy regimes could lead to long-term immunological control of HIV. *Proc. Natl. Acad. Sci. USA*, **96**, 14464–69.

Wodarz, D., Klenerman, P. and Nowak, M. A. (1998). Dynamics of cytotoxic T-lymphocyte exhaustion. *Proc. Roy. Soc. Lond. B*, **265**, 191–203.

Wodarz, D., Lloyd, A. L., Jansen, V. A. A and Nowak, M. A. (1999*a*). Dynamics of macrophage and T-cell infection by HIV. *J. Theor. Biol.*, **196**, 101–13.

Wodarz, D., Nowak, M. A. and Bangham, C. R. (1999*b*). The dynamics of HTLV-I and the CTL response. *Immunol. Today*, **20**, 220–7.

Wodarz, D., Hall, S. E., Usuku, K., Osame, M., Ogg, G. S., McMichael, A. J., Nowak, M. A. and Bangham, C. R. M. (2000*a*). CTL abundance and virus load in HIV-1 and HTLV-1. Preprint.

Wodarz, D., May, R. M. and Nowak, M. A. (2000*b*). The role of antigen-independent persistence of memory CTL. *Int. Immunol.*, in press.

Wolfs, T. F. W., Zwart, G., Bakker, M., Valk, M., Kuiken, C. L. and Goudsmit, J. (1991). Naturally occurring mutations within the HIV-1 V3 genomic RNA lead to antigenic variation dependent on a single amino-acid substitution. *Virology*, **185**, 195–205.

Wolinsky, S. M., Korber, B. T. M., Neumann, A. U., *et al.* (1996). Adaptive evolution of human immunodeficiency virus type-1 during the natural course of infection. *Science*, **272**, 537–42.

Wolthers, K. C., Noest, A. J., Otto, S. A., Miedema, F. and De Boer, R. J. (1999). Normal telomere lengths in naive and memory CD4+ T cells in HIV type 1 infection: a mathematical interpretation *AIDS Res. Hum. Retroviruses*, **15**, 1053–62.

Wong, J. K., Hezareh, M., Gunthard, H. F., *et al.* (1997). Recovery of replication-competent HIV despite prolonged suppression of plasma viraemia. *Science*, **278**, 1291–95.

Wong-Staal, F. and Gallo, R. C. (1985). Human T-lymphotropic retroviruses. *Nature*, **317**, 395–403.

Wong-Staal, F., Shaw, G. M., Hahn, B. H., Salahuddin, S. Z., Popovic, M., Markham, P., Redfield, R. and Gallo, R. C. (1985). Genomic diversity of human T-lymphotropic virus type III (HTLV-III). *Science*, **229**, 759–62.

Zagury, D., Bernard, J., Leonard, R., Cheynier, R., Feldman, M., Sarin, P. S. and Gallo, R. C. (1986). Long-term cultures of HTLV-III-infected T cells: a model of cytopathology of T-cell depletion in AIDS. *Science*, **231**, 850–3.

Zhang, L. Q., Mackenzie, P., Cleland, A., Holmes, E. C., Brown, A. J. L. and Simmonds, P. (1993). Selection for specific sequences in the external envelope protein of human immunodeficiency virus type-1 upon primary infection. *J. Virol.*, **67**, 1772–7.

Zinkernagel, R. M. (1996). Immunology taught by viruses. *Science*, **271**, 173–8.

Zinkernagel, R. M. and Hengartner, H. (1994). T-cell-mediated immunopathology versus direct cytolysis by virus—implications for HIV and AIDS. *Immunol. Today*, **15**, 262–8.

Zinkernagel, R. M. and Hengartner, H. (1997). Antiviral immunity. *Immunol. Today*, **18**, 258–60.

INDEX

Printed in the United States
By Bookmasters